DANTE
Lyric Poet and Philosopher

The geocentric universe of Pythagoras as depicted by Peter Apian in his *Cosmographia* of 1539. The Earth is shown surrounded by the four elements, the seven planetary spheres, the firmament or starry heaven (with zodiac), the crystalline heaven and *primum mobile* (for Dante one and the same ninth heaven), and the Empyrean.
Ann Ronan Picture Library.

DANTE

Lyric Poet and Philosopher

*An Introduction to the
Minor Works*

J. F. Took

CLARENDON PRESS · OXFORD
1990

Oxford University Press, Walton Street, Oxford OX2 6DP
Oxford New York Toronto
Delhi Bombay Calcutta Madras Karachi
Petaling Jaya Singapore Hong Kong Tokyo
Nairobi Dar es Salaam Cape Town
Melbourne Auckland
and associated companies in
Berlin Ibadan

Oxford is a trade mark of Oxford University Press

Published in the United States
by Oxford University Press, New York

© J. F. Took 1990

All rights reserved. No part of this publication may be reproduced, stored in a retrieval system, or transmitted, in any form or by any means, electronic, mechanical, photocopying, recording, or otherwise, without the prior permission of Oxford University Press

This book is sold subject to the condition that it shall not, by way of trade or otherwise, be lent, re-sold, hired out or otherwise circulated without the publisher's prior consent in any form of binding or cover other than that in which it is published and without a similar condition including this condition being imposed on the subsequent purchaser

British Library Cataloguing in Publication Data
Took, J. F.
Dante: lyric poet and philosopher: an introduction
to the minor works.
1. Poetry in Italian. Dante Alighieri, 1265–1321
I. Title
851.1
ISBN 0-19-815158-6

Library of Congress Cataloging in Publication Data
Took, John.
Dante, lyric poet and philosopher: an introduction
to the minor works / John Took.
p. cm.
Includes bibliographical references.
1. Dante Alighieri, 1265–1321—Criticism and interpretation.
I. Title.
PQ4308.T66 1990   851'.1—dc20   90–6888
ISBN 0-19-815158-6

Typeset by Pentacor plc, High Wycombe, Bucks.
Printed in Great Britain by Bookcraft (Bath) Ltd.
Midsomer Norton, Avon

For E., J., and P.
Ecce hereditas Domini filii (Ps. 127: 3)

Preface

DANTE'S, said the late Kenelm Foster, is a classic case of the masterpiece and the rest;[1] and in an obvious sense he was right. Indeed, one of the many things Kenelm Foster was able with his customary precision to demonstrate was the way in which two of the most important of the minor works—the *Convivio* and the late canzone *Tre donne intorno al cor*—are but outposts of the *Commedia*, distant harbingers of a greater maturity and equilibrium to come. But that is only half the story, for though the minor works are anticipatory with respect to the *Commedia*, they none the less have about them a characteristically Dantean magnanimity, a typically Dantean sense of the big issue. Take, for example, the *Vita nuova*. Ostensibly an account of Dante's meeting with Beatrice, it raises far-reaching issues of affective philosophy and literary aesthetics. By no means mere autobiography, it seeks to resolve a problem, to settle a long-standing difficulty. Or take the *Convivio*. Here too Kenelm Foster was right to see traces of immaturity, indeed of theological recklessness. But just think what Dante was trying to do. Heir to the Aristotelianism of the thirteenth century, but conscious of belonging to a special kind of communal, commercial, republican, and capitalistic civilization, he tried to bring the two together, to define a way of human happiness for those bowed down, as he put it (I. i. 13), 'by domestic and civic care'—and all this, not in Latin, but in Italian, in the language representative of the new society. It was a vast undertaking, and if it led to some distortion at the theological level, at the level of nature and grace, then so be it; that could be put right later. Or take the *De vulgari eloquentia*. Here is enterprise on a large scale; for Dante's aim now is to define a linguistic standard, an illustrious, cardinal, courtly, and curial tongue representative of the Italian people as a whole and embodying the substance and quality of their *italianitas*. Not content with actually doing the job, with (in the *Convivio*) fashioning from the young vernacular a language as capable expressively and aesthetically as Latin (*Conv*. I. x. 12), he wishes now to settle the matter theoretically, to nail it once and for all on a scientific basis. Or take the *Monarchia*, by any standards a remarkable foray into political theology, into the spelling out of God's plan for human government; no hint here of the mean-spirited or of niggling self-doubt. All instead is commitment, unqualified dedication to the grand design and to its forthright proclamation.

But that is not all, for the settling of the issue tends in Dante to take the form of confession, of a periodic enquiry into the state of his own soul; so, for example, the *Vita nuova*, an essay, certainly, in the philosophy of love and of literary beauty, but, in and through the philosophical, an exploration and presentation of his own experience in the received categories of the 'courtly' tradition, a marking out of positions reached and of truths embraced; so too the *Convivio*, the *De vulgari eloquentia*, and the *Monarchia*, each of them, for all its broad social conception, an act of intimate self-ordering, a tally of progress, a resolution of personality. All in all, then, the 'rest', in Dante, turns out to be quite a significant rest, a sizeable contribution to wisdom pure and simple, and an impressive espousal of the Augustinian notion of the text as confession. And that, basically, is the reason for this book. Conscious of the difficulty of these works, but also of their delight in the act of self-possession, I have tried to provide just enough background and commentary to move the reader on as quickly as possible to the original. Far from setting up another obstacle along the way, I have tried after Dante's own manner in the *Convivio* to provide a starter and stimulus to the main course.

There are four chapters. Chapter 1 has to do with the early *rime* up to the 'praise' poems inaugurated by the canzone *Donne ch'avete intelletto d'amore*, and with the *Vita nuova*. It includes too a discussion of the *Detto d'amore* and of the *Fiore*, which I take almost certainly to be Dante's and to predate the *Vita nuova* by four or five years. Chapter 2 begins with the mature *rime* and moves on to the *Convivio*, while Chapter 3 has to do with the *De vulgari eloquentia*. Chapter 4 focuses on the *Monarchia*, but includes towards the end an account of the letters, of the Latin treatise *De situ et forma aque et terre*, and of the eclogues. Background and general perspective are at each stage—though most conspicuously in the cases of the *Convivio* and of the *Monarchia*—complemented by a sense of the problematic, of what it is about these works that remains difficult and unsettled.

The notes and bibliography are intended to document the argument on the basis of texts most probably familiar to Dante, and to point the reader in the direction of further discussion. For reasons of space I have dispensed with translations of the original except in the case of Old French and of Provençal, where translation accompanies the text, and of Latin, where it stands in for the text. There is, however, a list of translations (where they exist) for the Italian in the bibliography.

My debts are considerable, and those among the myriad Dante scholars who have passed this way before and who have most influenced me I have tried to acknowledge in the bibliographies. I must, however, mention four in particular of my colleagues—Laura Lepschy, Conor Fahy, and Peter Armour of the University of London, and John Woodhouse of Magdalen College, Oxford—all of whom kindly read the typescript and suggested many valuable improvements. I am grateful too to the staff of Oxford University Press for their vigilance in the later stages of the book's production.

University College London, 1989   John Took

Contents

1	The Early *Rime*, the *Detto d'amore*, the *Fiore*, and the *Vita nuova*	1
	(i) The basic problem: love and language (Sicilians, Tuscans, and *stilnovisti*)	1
	(ii) Dante's early *rime*	12
	(iii) The *Detto d'amore* and the *Fiore*	29
	(iv) The *Vita nuova*	43
2	The Mature *Rime* and the *Convivio*	61
	(i) The *rime*	61
	(ii) The *Convivio*: introduction and course of the argument	81
	(iii) The problem of the *Convivio*: idealism and humanism	93
3	The *De vulgari eloquentia*	123
4	The *Monarchia*, the Letters, the *De situ*, and the Eclogues	147
	(i) The *Monarchia*: introduction and chronology	147
	(ii) Course of the argument	151
	(iii) Papal hierocracy, moral finalism, and the *Monarchia* as a work of Christian political philosophy	159
	(iv) The letters, the *De situ*, and the eclogues	173
Conclusion		188
Abbreviations		190
Notes		192
General Bibliography		221
Index		229

I

The Early *Rime*, the *Detto d'amore*, the *Fiore*, and the *Vita nuova*

(i)

The basic problem: love and language (Sicilians, Tuscans, and stilnovisti)

DANTE, whatever else he was, was a great organizer. He loved nothing more than to pause, to dismantle his experience so far, to examine its component parts, to discard the no longer efficient, to put it all back together, and to admire its new-found harmony and proportion. The *Vita nuova* is exemplary. Conceived after the death of Beatrice and in anticipation of a new kind of literary and speculative activity, it sets out to trace the course of his experience so far as a philosopher and poet of love, and to demonstrate its essential unity. Far from a mere anthology or collection of juvenilia with commentary, it testifies to an order of experience—'neo-courtly' we might call it[1]—passionately lived and subject now to systematic statement. But the resolving of his own experience meant for Dante the resolving of his tradition too, the reaching out through confession to settle a public issue. Here, then, with the public dimension of the book, with the prior and long-standing problems of love and of style, is where we shall begin.

The problem of love derived ultimately from Ovid's *Ars amatoria*. Not that the *Ars* itself, a manual for seducers, was problematic. What was problematic was its reception in the Christian-courtly circles of medieval France and Provence;[2] for love, though still a sensual experience, took on now, in keeping with its new circumstances, a salutary aspect. It was good for the soul, morally improving. Andreas Capellanus, author of the late twelfth-century *De amore*, puts it well.[3] Love, he has one of his interlocutors say, is what it always was: carnal. First comes the giving of hope, then the kiss, then the embrace, and finally the 'yielding of the whole person' (I. vi; ed. Walsh, 57). At the same time, however, it is educative, apt to make the rough man smooth, the humble man noble, the proud man humble, and to fill everyone with the spirit of beneficence:

> The effect of love is that no greed can cheapen the true lover. Love makes the hirsute barbarian as handsome as can be: it can even enrich the lowest-born with nobility of manners: usually it even endows with humility the arrogant. A person in love grows to the practice of performing numerous services becomingly to all. What a remarkable thing is love, for it invests a man with such shining virtues, and there is no one whom it does not instruct to have these great and good habits in plenty!
>
> (I. iv; ed. Walsh, 39)

Now it is difficult not to sense in all this a wisp of comic irony, of the self-conscious delight in love's paradoxes flowing down through Guillaume de Lorris and Jean de Meun of the *Roman de la rose* to and beyond Durante—probably our own Dante—of the *Fiore*. But irony was not the only way of handling love's ambiguity. Such was the complexity of the question that there was room too for the psychologists and moralists, for those anxious (*a*) to investigate the structure of amorous experience in and for itself, and (*b*) to stress love's power to reform spiritually. Characteristic, therefore, of the Sicilians, the first generation of Italian 'courtly' poets (*fl.* 1230–66),[4] is their interest, not in the contingencies of love, in its time, place, and circumstances, but in its condition, in the experience of seeing, sorrowing, sighing, and rejoicing. True, lovers' meetings, partings, and debates (*contrasti*) still figure prominently, but by and large the anecdotal serves the analytical, the poem's basic rhythm of introspection; so, for example, the Notaro Giacomo da Lentini, content certainly to work with the themes, rhetorical techniques, and images of his Provençal predecessors, but fashioning even so a line nicely responsive to (in this case) love's pain and yearning:

> Lo vostr'amor che m'ave
> in mare tempestoso,
> è sì como la nave
> c'a fortuna getta ogni pesanti,
>   e campan per lo getto
> di loco periglioso.

> Similemente eo getto
> a voi, bella, li mei sospiri e pianti:
>   ché, s'eo no li gittasse,
> parria che soffondasse
> (e bene soffondara)
> lo cor, tanto gravara—in suo disio;
>   ché tanto frange a terra
> tempesta, che s'aterra;
> ed eo così rinfrango:
> quando sospiro e piango,—posar crio.
>
>   (*Madonna, dir vo voglio*, lines
>   49–64; ed. Contini, i. 53)

The Tuscans, by contrast, mainland successors to the Sicilians, took a moral line.[5] Mildly discomforted by Sicilian hedonism, they tended to gloss the whole thing communally, in keeping with the exigencies of the bourgeois conscience. Love was a good thing, liable in the right-thinking citizen to inculcate the virtues of patience, perseverance, and self-discipline. So, for example, Bonagiunta da Lucca (d. after 1296):

> Fin'Amor mi conforta
> e lo cor m'intalenta
> madonna, ch'io non penta
> di voi s'io innamorai.
>   Membrando ciò che porta
> la vita n'è contenta,
> avegna ch'io ne senta
> tormenti pur assai.
>   Ca primamente amai
> per ben piacere al vostro signoraggio
> d'aver fermo coraggio,
> a ciò ch'io per fermeze non dottasse
> che 'l meo lavor falsasse;
> ché ch'incomenza mez'ha compimento
> se sa perseverare lo suo adoperamento.
>   Ed io perseverando . . .
>   (the second stanza goes on)
>
>   (*Fin'Amor mi conforta*, lines 1–16;
>   ed. Salinari, 290)

and, Tuscan-in-chief, Guittone d'Arezzo (d. *c.* 1294):[6]

> Om ch'ama pregio e pò,
> più che legger en scola,
> Amor valeli pro:
> ché più leggero è Po
> a passar senza scola
> che lo mondo a om pro'
> senza Amor, che dà
> cor e bisogno da
> sprovar valor e forzo;
> perché ciascun om, for zo
> che briga e travagli' agia,
> se vale, non varrà già.
>
> (*Tuttor, s'eo veglio o dormo*, lines 25–36; ed. Contini, i. 198)

But these, interesting as they are as emphases, as responses evoked by this or that set of social and intellectual circumstances to the phenomenon of 'courtly love', *are* emphases. They are not solutions, the solving of love's ambiguity falling, not to the Sicilians or Tuscans, but to the *stilnovisti*, to the small group of Emilian and Tuscan poets consisting of (together with a number of *minori*) Guido Guinizzelli (d. *c.*1276), Guido Cavalcanti (d. 1300), Dante, and Cino da Pistoia (d. 1337).[7] Only now, on the basis of an appeal (*a*) to the metaphysic and psychology of the new Aristotelianism, and (*b*) to Christian (though originally Stoic) literature *de amicitia spirituali* (on spiritual friendship), was love redefined as a matter less of possession than of 'becoming', of 'openness to' the good over and beyond self. Only now did it assume the aspect of disposition as distinct from acquisition. Here, though, the question becomes complicated, for the *stilnovisti*, united as they were in stressing love's *interiorità*, its new and unashamedly superior inwardness, were far from agreed as to the details. The odd man out was Cavalcanti, who, ruthless in his indictment of Tuscan neo-Ovidianism—of the failure of the Guittoniani to break with the old way of thinking—none the less saw love as irrational, indeed as destructive.[8] The key text is his difficult canzone *Donna me prega*, five stanzas (plus *congedo*) of the densest scholastic argumentation. Love, he says, though generated by visual perception and holding out the prospect of contemplative peace, is in reality a dark, restive, and malign passion of the sensitive soul. It affords the lover no respite. It waylays him judgementally—

> for di salute—giudicar mantene,
> ché la 'ntenzione—per ragione—vale:
> discerne male—in cui è vizio amico.
>
> (lines 32–4; ed. Contini, ii. 526)

—and hastens his spiritual demise:

> Di sua potenza segue spesso morte,
> se forte—la vertù fosse impedita,
> la quale aita—la contraria via:
> non perché oppost'a naturale sia;
>   ma quanto che da buon perfetto tort'è
> per sorte,—non pò dire om ch'aggia vita,
> ché stabilita—non ha segnoria.
>
> (lines 35–41; ibid.)

Hence the tragedy of Cavalcanti's poetic world, its pervasive sense of conflict. On the one hand stands the possibility of understanding, of abstraction, imagination, and contemplation issuing in a state of pure consciousness (eloquent here, for example, is the 'praise poem' *Chi è questa che vèn, ch'ogn'om la mira*); on the other hand stands love, invasive, ineluctable, and inducing in the lover a deep sense of helplessness as he witnesses his own spiritual disintegration:

> L'anima mia vilment'è sbigotita
> de la battaglia ch'ell'ave dal core:
> che s'ella sente pur un poco Amore
> più presso a lui che non sòle, ella more.
>
> Sta come quella che non ha valore,
> ch'è per temenza da lo cor partita;
> e chi vedesse com'ell'è fuggita
> diria per certo: 'Questi non ha vita.'
>
> Per li occhi venne la battaglia in pria,
> che ruppe ogni valore immantenente,
> sì che del colpo fu strutta la mente.
>
> Qualunqu'è quei che più allegrezza sente,
> se vedesse li spirti fuggir via,
> di grande sua pietate piangeria.
>
> (ibid. ii. 498)

It was only with Guinizzelli, then—though Guinizzelli, strictly, predates Cavalcanti—that the new and more characteristically stilnovistic sense of love as *becoming*, as self-transcendence through contemplation, was made secure.[9] Guinizzelli started out a Guittonian, his (presumably early) canzoni *Tegno de folle 'mpres', a lo ver dire* and *Madonna, il fino amor ched eo vo porto* being rich in every kind of Provençal strategy from *capfinidas*—the linking of stanzas by the same word—down to the details of lexis ('sclarisce', 'enveggia', 'clarore', etc.) and grammatical inflexion ('-anza', '-enza'). More congenial, however, was the 'visionary' mood of the Sicilians, which served to encourage him in the recognizably stilnovistic register of *Gentil donzella, di pregio nomata* and of the exquisite *Io voglio del ver la mia donna laudare*, a replica—but marvellously fashioned—of Giacomo da Lentini's *Madonna ha 'n sé vertute con valore*:

> Io voglio del ver la mia donna laudare
> ed asembrarli la rosa e lo giglio:
> più che stella dïana splende e pare,
> e ciò ch'è lassù bello a lei somiglio.
>
> Verde river' a lei rasembro e l'âre,
> tutti color di fior', giano e vermiglio,
> oro ed azzurro e ricche gioi per dare:
> medesmo Amor per lei rafina meglio.
>
> Passa per via adorna, e sì gentile
> ch'abassa orgoglio a cui dona salute,
> e fa 'l de nostra fé se non la crede;
>
> e no.lle pò apressare om che sia vile;
> ancor ve dirò c'ha maggior vertute:
> null'om pò mal pensar fin che la vede.
>
> (ed. Contini, ii. 472)

This is the context of the great manifesto canzone *Al cor gentil rempaira sempre amore*, a curious combination of grandeur and inhibition, of novel intuition and old-fashioned 'courtly' prejudice.[10] On the one hand, Guinizzelli says, love and the noble soul ('cor gentil') belong together as the bird and the wood or as warmth and firelight. The one, the latter, is the precondition of the other:

> Al cor gentil rempaira sempre amore
> come l'ausello in selva a la verdura;
>   né fe' amor anti che gentil core,
> né gentil core anti ch'amor, natura:
>   ch'adesso con' fu 'l sole,
> sì tosto lo splendore fu lucente,
> né fu davanti 'l sole;
> e prende amore in gentilezza loco
> così proprïamente
> come calore in clarità di foco.
>
>   (lines 1–10; ed. Contini, ii. 460–1)

But nobility, though a *principle* of love, is at the same time a *product* of love; for just as the Intelligences are moved through contemplation to obey their Maker, so also the lover is moved through contemplation to obey his lady, this being the way of spiritual renewal:

> Splende 'n la 'ntelligenzïa del cielo
> Deo crïator più che 'n nostr'occhi 'l sole:
>   ella intende suo fattor oltra 'l cielo,
> e 'l ciel volgiando, a Lui obedir tole;
>   e con' segue, al primero,
> del giusto Deo beato compimento,
> così dar dovria, al vero,
> la bella donna, poi che 'n gli occhi splende
> del suo gentil, talento
> che mai di lei obedir non si disprende.
>
>   (lines 41–50; ibid. ii. 463)

The parallel is breath-taking, and rich (as Dante for one immediately saw) in every kind of ontological and moral possibility. *Madonna* is a 'fattore',[11] a *maker*, in the moral order, while the lover, through obedience to her, is made over anew, reconstituted in the depths of his being. In fact, Guinizzelli immediately takes with one hand what he has just given with the other ('I did not mean to idolize my lady,' he says in the last stanza, 'it was simply that she *seemed* so angelic'); hence the inhibition and provisionality of his canzone, its stopping short of the genuinely innovative. But that, for Dante, did not matter. What mattered was its potential, the basis it offered for an altogether bolder solution to the problem of love.

The problem of style too goes back a long way, as far this time as the great rhetorical treatises of classical antiquity. The point is that rhetoric as classically conceived involves typically a distinction between the *what* and the *how* of literature, between what is said and the way it is said. The various figures of thought and speech as listed by the manuals—by the pseudo-Ciceronian *Rhetorica ad Herennium*, for example, or by the *Institutio oratoria* of Quintilian—are 'applied', added on by way of technical elevation or *exornatio*. Take, for example, the following passages from the *Ad Herennium* (IV. viii. 11) and from the *Institutio oratoria* (VIII. iii. 61) respectively:

A discourse will be composed in the Grand style if to each idea are applied the most ornate words that can be found for it, whether literal or figurative.

The ornate style is something that goes beyond what is merely lucid and acceptable. It consists firstly in forming a clear conception of what we wish to say, secondly in giving this adequate expression, and thirdly in lending it additional brilliance, a process which may correctly be termed embellishment.

True, the *what* and the *how* ought to bear some relationship, and prominent in the rhetorical manuals both of antiquity and of the Middle Ages is a sense of the *convenience* or decorum by virtue of which form is properly matched to substance, properly constrained by the nature and tone of what is being said. But that, important as it is, does not alter the basic sense of rhetoric as *additio*, as something added on in a subsequent phase of technical elaboration.

Now the influence of classical rhetoric on the vernacular lyric poetry of France, Provence, and (up at least to Dante's time) Italy is hard to assess.[12] What is clear, however, is that this is an order of verse in the highest degree stylistically aware. When, for example, a Pierre d'Alvernh, or a Giraut de Bornelh, or an Arnaut Daniel announces, as frequently he does, that he will sing in an 'open' (*leu*) style or in a 'closed' (*clus*) and 'difficult' style, he is announcing that style constitutes part of the poem's substance. The *how* is part of the *what*. And the same applies in a narrower and more selective way to the Sicilians, who found in the *trobar leu* or 'accessible' style of the Provençal poets a register congenial to their psychological interests. The result was a line marked at its best by a distinctive purity and suppleness, by a lexis and syntax responsive in their 'openness' to the ease and transparency of the poets' inspiration. The Tuscans by contrast, seeking in the way we have seen to gloss the love lyric morally, succeeded on the stylistic level merely in complicating it, in imposing on it a conscientious but indigestible rhetoricism. There is, it is true, a certain authenticity about the rhetoricism of the Tuscans, the authenticity of civic pride and, in the case of Guittone, of moral anguish. Its knottiness is to an extent the knottiness of the problem it addresses, of how to resolve the by now infinitely complex nucleus of 'courtly' themes in terms acceptable, indeed intelligible, to the bourgeois mentality. But that was not how it seemed at the time. At the time it seemed a matter of difficulty for difficulty's sake. What, for example, was a sophisticated generation of philosopher-poets such as the *stilnovisti*, like the Sicilians exquisitely sensitive to the purity and musicality of the line, to make of pieces like the following (one of Guittone's)?

> Già lungiamente sono stato punto,
> sì punto—m'ave la noiosa gente,
> dicendo de saver uve mi punto;
> sì tal punto—mi fa quasi piangente.
>
> Poi, se.mmi miro, non credone punto,
> sì punto—so', 've 'n stando onor v'è gente,
> poi lo mïo voler de gioi' ha punto,
> che punto—è verso, sì face ha piagente.
>
> Ferò como lo bono arcero face:
> face—fa de fedire in tale parte,
> sparte—di ciò, u' non par badi, fede.
>
> A tutti amanti sì de' farse face:
> sface—ciò de penser l'aversa parte,
> parte—che vive in error de su' fede.
>
> (ed. Contini, i. 249)

Not much. Totally muscle-bound in its repetition and equivocation, in its sheer virtuosity, the sonnet was little short of offensive. This, then, was the burden of the stilnovistic critique of Guittone and of the Guittoniani generally—that their poetry lacked any sense of finesse. There are two documents in particular to consider. The first is Cavalcanti's sonnet, part of an exchange with Guido Orlandi, *Di vil matera mi conven parlare*. Spiritedly, not to say tetchily, Cavalcanti vindicates over and against the outmoded Ovidianism of his Guittonian colleagues a fresh sense of love as 'discursiveness' and of style as faithfulness, as responsiveness to what Dante calls its 'inner dictation' or 'dittar dentro':

> Di vil matera mi conven parlare,
> e perder rime, silabe e sonetto,
> sì ch'a me stesso giuro ed imprometto
> a tal voler per modo legge dare.
>
> Perché sacciate balestra legare
> e coglier con isquadra archile in tetto,
> e certe fiate aggiate Ovidio letto,
> e trar quadrelli e false rime usare,
>
> non pò venire per la vostra mente
> là dove insegna Amor, sottile e piano,
> di sua manera dire e di su' stato.
>
> Già non è cosa che si porti in mano:
> qual che voi siate, egli è d'un'altra gente:
> sol al parlar si vede chi v'è stato.
>
> Già non vi toccò lo sonetto primo:
> Amore ha fabricato ciò ch'io limo.
>
> (ibid. ii. 563–4)

Because you have read Ovid, he says, and because you can concoct the odd clever rhyme, do not for one moment think you understand love; for love, far from a matter of display, is one of sweet 'insegnamento', of subtle intuition and of acquiescence in the good beyond self. And style, far from a matter of virtuosity, is one of 'obedience', of hearkening to the movement of conscience and of refining the word accordingly ('Amore ha fabricato ciò ch'io limo'). You can always tell, Cavalcanti says, who has been there and who has not; and you, he tells Orlandi, have not ('qual che voi siate, egli è d'un'altra gente'). Sharply, then, cruelly even, but with a typical sense of the triumph of the new over the old, Cavalcanti makes explicit the leading principles of his and of the stilnovistic aesthetic generally: limpidity, grace, discretion, musicality—a clarity, in short, of style, syntax, and sound in keeping with the delicate movement of awareness.[13]

The other document is the retrospective synthesis of *Purgatorio* XXIV. 49–63. Here, admittedly, we are on different ground, for Dante's words have to be interpreted not only, nor even primarily, in the context of the original polemic of twenty or thirty years earlier, but of the new spirituality of the *Commedia*. The whole question of poetry and of poetics is bound up now with that of confession, repentance, and, in a specifically Christian sense, of purgation. Redefined as it is, however, the issue remains the same; here as before it is a question of correspondence, of the proper enmeshing of form and substance in the light of right understanding. The passage runs as follows:

> 'Ma dì s'i' veggio qui colui che fore
> trasse le nove rime, cominciando
> *Donne ch'avete intelletto d'amore.*'
>   E io a lui: 'I' mi son un che, quando
> Amor mi spira, noto, e a quel modo
> ch'e' ditta dentro vo significando.'
>   'O frate, issa vegg'io', diss'elli, 'il nodo
> che 'l Notaro e Guittone e me ritenne
> di qua dal dolce stil novo ch'i' odo!
> Io veggio ben come le vostre penne
> di retro al dittator sen vanno strette,
> che de le nostre certo non avvenne;
>   e qual più a gradire oltre si mette,
> non vede più da l'uno a l'altro stilo';
> e, quasi contentato, si tacette.

As in Cavalcanti, the new, both ideologically and stylistically, is affirmed over and against the old (represented here by Bonagiunta). Ideologically, it is a question of love as sweet instruction, as the movement of thought and feeling as perceived and enjoyed in the forum of consciousness. Stylistically, it is a matter, not of addition or of imposition, but of giving shape 'scribally' to the otherwise hidden and inscrutable 'dictation within'. On both counts, the old is done away with. Love, far from possession, is discourse, awareness, new life; and style, far from decoration, is reasonable self-affirmation, the means and condition of intimate understanding, indeed love itself under the aspect of intelligibility.

(ii)
Dante's early rime

It is sometimes said that the first real, as distinct from merely derived, school of Italian lyric art was the *stil novo*, and further that it was as a *stilnovista* that Dante discovered his authentic voice as a poet. Here for the first time he accomplished his ideal union *in re*, in the poem, of form and substance, his ideal accountability of technique to the movement of thought and emotion (the 'dittar dentro'). Up till then, all had been experimentation and ambition, experimentation conducted under the influence of Guittone, of Cavalcanti, and of Guinizzelli, and ambition on the part of the young poet to make his name. And that, as far as it goes, is fine. There is no denying in the earliest extant poems of Dante their blend of conformity and competitiveness, of submissiveness and, especially on the technical front, flamboyant self-assertion. But there is more, for Dante's years as an apprentice are marked by a desire, not simply for recognition within the dominant literary circles, nor even for technical mastery, but to press the possibilities of his tradition to the limit, to elicit and to exhaust its conceptual and formal content. Right from the start we sense something of his 'serietà terribile',[14] of his desire to establish the poem at the centre of moral and existential interest.

Typical are the early exchanges with Dante da Maiano, a conservative poet in the old Tuscan school.[15] Each testifies to the younger poet's impatience with received literary-ideological solutions and to a desire to reinforce their moral and psychological substance. The first begins with an enigma sonnet offered by Dante da Maiano for interpretation (*Provedi, saggio, ad esta visïone*). A scherzo in the line of the Provençal *partimen* or literary debate, it is rich in every sort of Gallic and Provençal archaism, and its levity of inspiration ('del più non dico, ché mi fe' giurare', line 13) is barely disguised. It is, in short, a bagatelle, the work of one content to 'trovare' in the time-honoured way, to stop short of true engagement. Dante's reply too is archaic. Conscious of his correspondent's elevated manner, he replies in kind ('com so rispondo a le parole ornate', *Savete giudicar vostra ragione*, line 4), and with a conventional deference to his partner's wisdom in matters of love. But immediately evident is a new seriousness of purpose. Wholly in contrast with Dante da Maiano's eroticism, Alighieri's ponderous hendecasyllables turn about the concepts of *valore*, *bieltate*, *amore*, and *fermezza*, with attention at once directed away from the game towards a more rigorous sense of love as a principle of moral and intellectual being:

> Disio verace, u' rado fin si pone,
> che mosse di valore o di bieltate,
> imagina l'amica oppinïone
> significasse il don che pria narrate
>
> . . . . .
>
> La figura che già morta sorvene
> è la fermezza ch'averà nel core.
> (*Savete giudicar vostra ragione*,
> lines 5–8, 13–14)

Already, then, there is in Dante a movement away from the dilettantism and social pleasantry of the old-style Tuscan sonneteer. There is a new philosophical urgency, a need to define, to conceptualize, to trace a psychological pattern, to intensify.

The same goes for the *duol d'amore* exchange, again with Dante da Maiano. The question proposed by Dante (assuming that it *is* Dante who opens this exchange—there is here a problem of attribution) relates to love's greatest sorrow: what, Dante asks, is love's greatest sorrow? To love and not to be loved in return, says Dante da Maiano, sorting through his repertory; this is love's greatest sorrow. But Alighieri is not satisfied. Struck by his correspondent's predictability, he presses for something further, something morally and intellectually more substantial ('e manti dicon che più v'ha dol maggio', *Lo vostro fermo dir*, line 11). But Dante da Maiano is finished. His imagination is spent. The best he can do is to make good rhetorically, by way of the headlong *rime equivoche* and *composte* of *Non canoscendo, amico, vostro nomo* ('par l'à / par là / parla'; 'nomo / d'un omo'; 'chi ama / chiama'; 'ch'amato / camato'), to which Alighieri, by no means to be outdone, replies in kind with *Lasso, lo dol che più mi dole e serra*. But what matters here is not so much the fact of technical sophistication—a second and post-stilnovistic phase of Guittonianism will see these same techniques justified intrinsically, in point of true accountability to the spirit's 'dittar dentro'—as Dante's restlessness, his desire to breach the conventional and the commonplace in favour of something dialectically more energetic:

> Però pregh'eo ch'argomentiate, saggio,
> d'autorità mostrando ciò che porta
> di voi la 'mpresa, acciò che sia più chiara;
> e poi parrà, parlando di ciò, chiara,
> quale più chiarirem dol pena porta,
> d'ello assegnando, amico, prov'e saggio.
>
> (*Lasso, lo dol che più mi dole e serra*,
> lines 9–14)

Finally, as far as Dante da Maiano is concerned, there is the exchange beginning *Amor mi fa sì fedelmente amare*, again to do with unrequited love. And again Dante da Maiano moves comfortably within the sphere of literary commonplaces, concluding with the familiar notion that service and long-suffering are all. Doesn't Dante agree? Dante's reply, if by 'reply' is meant an attempt to meet the enquirer on his own ground, is hardly a reply at all. Acknowledging as best he can his correspondent's presence ('Onde se voli, amico . . .'), he at once dismisses patience and long-suffering in favour of wisdom, courtesy, wit, art, nobility, beauty, courage, humility, magnanimity, resolve, and rectitude as the means of winning love over. Here too, then, the whole question is shifted on to another plane. The tired formulas of tradition are replaced by a fresh meditation on the moral, social, and aesthetic aspects of love's working out in the life and experience of the lover, by something intrinsically more purposeful:

> Savere e cortesia, ingegno ed arte,
> nobilitate, bellezza e riccore,
> fortezza e umiltate e largo core,
> prodezza ed eccellenza, giunte e sparte,
>
> este grazie e vertuti in onne parte
> con lo piacer di lor vincono Amore:
> una più ch'altra ben ha più valore
> inverso lui, ma ciascuna n'ha parte.
>
> Onde se voli, amico, che ti vaglia

> vertute naturale od accidente,
> con lealtà in piacer d'Amor l'adovra,
> e non a contastar sua graziosa ovra:
> ché nulla cosa gli è incontro possente,
> volendo prender om con lui battaglia.

Belonging more or less to the same period as the early exchanges with Dante da Maiano is a group of Siculo-Tuscan love poems consisting of poems 2–5 of the *Vita nuova* and, among those excluded from the *Vita nuova*, of the canzone stanzas *Lo meo servente core* and *Madonna, quel signor che voi portate*, the canzone *La dispietata mente*, and the sonnets *Volgete li occhi*, *Deh, ragioniamo insieme*, and *Com più vi fere Amor co' suoi vincastri*. Thematically, these poems are retrospective in their sense of love as a matter of solicitation and reward, this in turn giving rise to a certain imaginative inconsistency and rhetorical inertia; so, for example, the following stanza from *La dispietata mente*, with its hurried amalgam of 'courtly' and scriptural motifs (the 'supplication' motif of lines 14–16, the 'worthy servant' motif of lines 17–19, the 'portrait within' motif of lines 20–2, and, a shade incongruously, the 'created in God's image' motif of lines 23–6):

> Piacciavi, donna mia, non venir meno
> a questo punto al cor che tanto v'ama,
> poi sol da voi lo suo soccorso attende;
>   ché buon signor già non ristringe freno
> per soccorrer lo servo quando 'l chiama,
> ché non pur lui, ma suo onor difende.
>   E certo la sua doglia più m'incende,
> quand'i' mi penso ben, donna, che vui
> per man d'Amor là entro pinta sete:
>   così e voi dovete
> vie maggiormente aver cura di lui;
> ché Que' da cui convien che 'l ben s'appari,
> per l'imagine sua ne tien più cari.

(lines 14–26)

or the following lines (13–20) from *O voi che per la via d'Amor passate* (*Vn* VII), with their accumulation of *conversiones* (the substitution of nominal for verbal or for adjectival forms of expression) and binary repetition:

> Or ho perduta tutta mia *baldanza*,
> che si movea d'amoroso tesoro;
> ond'io pover dimoro,
> in guisa che di dir mi ven *dottanza*.
>
> Sì che volendo far come coloro
> che per vergogna celan lor *mancanza*,
> di fuor mostro *allegranza*,
> e dentro da lo core *struggo e ploro*.

The criticism, it is true, has not to be pressed too hard, since stylistically these Siculo-Tuscan love poems (to which should be added the *planctus* or weeping poem *Piangete, amanti* and the *vituperatio* or accusation poem *Morte villana* in Chapter VIII of the *Vita nuova*) reflect the easier or 'open' (*leu*) style rather than the more difficult *clus* style of Tuscan lyricism, the style of Bonagiunta rather than that of Guittone. *Deh, ragioniamo*, moreover, has something of the madrigalesque about it, something of the exquisite *Un dì si venne a me Malinconia* to come:

> Deh, ragioniamo insieme un poco, Amore,
> e tra'mi d'ira, che mi fa pensare;
> e se vuol l'un de l'altro dilettare,
> trattiam di nostra donna omai, signore.
>
> Certo il vïaggio ne parrà minore
> prendendo un così dolze tranquillare,
> e già mi par gioioso il ritornare,
> audendo dire e dir di suo valore.
>
> Or incomincia, Amor, ché si convene,
> e moviti a far ciò ch'è la cagione
> che ti dichini a farmi compagnia,
>
> o vuol merzede o vuol tua cortesia;
> ché la mia mente il mio penser dipone,
> cotal disio de l'ascoltar mi vene.

But even this, for all its deftness of touch, progresses no further in point of understanding than *La dispietata mente*, the group as a whole, therefore, representing an early, pre-Cavalcantian phase of Dante's activity as a lyric poet.

Cavalcanti's, in fact, was the decisive influence on Dante, his true point of departure. Under it, he embarked on a new and meticulous poetic discipline designed to put an end to, or rather intimately to renew, the old conceptual and technical clichés of the 'courtly' tradition. Conceptually, it was a question as in Cavalcanti of love conceived tragically, as conducive in its irrationality, not to the lover's well-being, but to his hasty demise; so, for example, the *Vita nuova* sonnet *Spesse fiate vegnonmi a la mente*:

> Spesse fiate vegnonmi a la mente
> le oscure qualità ch'Amor mi dona,
> e venmene pietà, sì che sovente
> io dico: 'Lasso!, avviene elli a persona?';
>
> ch'Amor m'assale subitanamente,
> sì che la vita quasi m'abbandona:
> campami un spirto vivo solamente,
> e que' riman perché di voi ragiona.
>
> Poscia mi sforzo, ché mi voglio atare;
> e così smorto, d'onne valor voto,
> vegno a vedervi, credendo guerire:
>
> e se io levo li occhi per guardare,
> nel cor mi si comincia uno tremoto,
> che fa de' polsi l'anima partire.

where, with a psychological finesse and dramatic immediacy wholly Cavalcantian in spirit, the lover comments on the perplexity and fear brought about by love's relentless assault on his soul. This, though, was quickly to be superseded in Dante by a Guinizzellian sense of love as spiritually reconstitutive, with the result that what matters about *Spesse fiate* is not so much its substance as its form, its sense of the word, not as 'additional', but as analytical, as a means, not to the embellishment, but to the establishment of experience. Gone are the conversions, the repetitions, and the stylistic strategies of the Guittoniani. The 'instance' or rhetorical flourish gives way to syntax ('e venmene pietà, sì che sovente . . . ch'Amor m'assale . . . sì che la vita . . . Poscia mi sforzo, ché mi voglio atare . . . e così smorto . . . e se io levo . . . che fa de' polsi') as the poet's chief expressive means. So too, in addition to *Spesse fiate*, the *Vita nuova* sonnet *Ciò che m'incontra*:

> Ciò che m'incontra, ne la mente more,
> quand'i' vegno a veder voi, bella gioia;
> e quand'io vi son presso, i' sento Amore

> che dice: 'Fuggi, se 'l perir t'è noia.'
>
> Lo viso mostra lo color del core,
> che, tramortendo, ovunque pò s'appoia;
> e per la ebrietà del gran tremore
> le pietre par che gridin: Moia, moia.
>
> Peccato face chi allora mi vide,
> se l'alma sbigottita non conforta,
> sol dimostrando che di me li doglia,
>
> per la pietà, che 'l vostro gabbo ancide,
> la qual si cria ne la vista morta
> de li occhi, c'hanno di lor morte voglia.

where again the discursive prevails over the elaborative, the analytical over the rhetorically florid. True, *Lo doloroso amor* and *E' m'incresce di me* are stylistically more ambiguous, more wedded despite their Cavalcantian inspiration to Tuscan mannerism.[16] But in the *Vita nuova* sonnets, subsequently chosen by Dante as representative of his progress as a technician, there is a fresh poetic integrity, a sharper sense of form in relation to content.

Here, though, we need to pause, for important in parenthesis are those poems reflecting the other side of Cavalcanti, Cavalcanti of the Sicilianizing ballads *Fresca rosa novella*, *Perch'i' no spero di tornar giammai*, and *Era in penser d'amor*. These, as examples of the genre, deserve the closest attention, for they are masterpieces of tact and musicality; so, for example, the delightful *Per una ghirlandetta*, where a neat interlacing of *settenari* and *novenari* combines on the imaginative plane with a delicate hint of the rococo to produce a moment of pure fantasy:

> Per una ghirlandetta
> ch'io vidi, mi farà
> sospirare ogni fiore.
>
> I' vidi a voi, donna, portare
> ghirlandetta di fior gentile,
> e sovr'a lei vidi volare
> un angiolel d'amore umile;
> e 'n suo cantar sottile
> dicea: 'Chi mi vedrà
> lauderà 'l mio signore.'
>
> Se io sarò là dove sia

> Fioretta mia bella a sentire,
>   allor dirò la donna mia
> che port' in testa i miei sospire.
>   Ma per crescer disire
> mïa donna verrà
> coronata da Amore.
>
>   Le parolette mie novelle,
> che di fiori fatto han ballata,
>   per leggiadria ci hanno tolt'elle
> una vesta ch'altrui fu data:
>   però siate pregata,
> qual uom la canterà,
> che li facciate onore.

—or the ballad *Deh, Vïoletta, che in ombra d'Amore*:

> Deh, Vïoletta, che in ombra d'Amore
> ne gli occhi miei sì subito apparisti,
> aggi pietà del cor che tu feristi,
> che spera in te e disïando more.
>
>   Tu, Vïoletta, in forma più che umana,
> foco mettesti dentro in la mia mente
>   col tuo piacer ch'io vidi;
>   poi con atto di spirito cocente
> creasti speme, che in parte mi sana
>   là dove tu mi ridi.
>   Deh, non guardare perché a lei mi fidi,
> ma drizza li occhi al gran disio che m'arde,
> ché mille donne già per esser tarde
> sentiron pena de l'altrui dolore.

—less satisfactory, it is true, for its oscillation between the popular ('aggi pietà del cor che tu feristi') and the esoteric ('Tu, Vïoletta, in forma più che umana'), but exemplary even so as an essay in Cavalcantian mannerism.

More remarkable still, however, or at least if not more remarkable then more celebrated, is the *locus amoenus* sonnet *Guido, i' vorrei che tu e Lapo ed io*, to which we might add the hunting sonnet—more in the vein, though, of Folgore da San Gimignano than of Cavalcanti—*Sonar bracchetti, e cacciatori aizzare*. *Guido, i' vorrei* is a gem of poetic discretion.[17] Conceived in the tradition of the Provençal *plazer* or 'pleasure' poem, but dominated by the idea of mystic quest and communion traceable possibly to the chivalric literature of France and Brittany, it is a confession of faith in the new poetic. Set adrift in a little boat, Dante and his two companions, Cavalcanti and co-*stilnovista* Lapo Gianni, will meditate on the experience and meaning of love, celebrated by them in verses, as here, of limpid grace and musicality:

> Guido, i' vorrei che tu e Lapo ed io
> fossimo presi per incantamento,
> e messi in un vasel ch'ad ogni vento
> per mare andasse al voler vostro e mio,
>
> sì che fortuna od altro tempo rio
> non ci potesse dare impedimento,
> anzi, vivendo sempre in un talento,
> di stare insieme crescesse 'l disio.
>
> E monna Vanna e monna Lagia poi
> con quella ch'è sul numer de le trenta
> con noi ponesse il buon incantatore:
>
> e quivi ragionar sempre d'amore,
> e ciascuna di lor fosse contenta,
> sì come i' credo che saremmo noi.

Sonar bracchetti, by contrast, moves from the static to the dynamic, from contemplation to activity. Dante takes himself to task for deserting the ladies for sport, for the belling of the hounds, for the starting of the hares, and for the headlong rush of the hunt through fair meadows. The social tableau is finely conceived and executed. But what matters again is the delicate turn of Dante's imagination, his superb deftness of touch in fashioning a sonnet as neat technically as it is inspirationally:

> Sonar bracchetti, e cacciatori aizzare,
> lepri levare, ed isgridar le genti,
> e di guinzagli uscir veltri correnti,
> per belle piagge volgere e imboccare,
>
> assai credo che deggia dilettare
> libero core e van d'intendimenti!
> Ed io, fra gli amorosi pensamenti,
> d'uno sono schernito in tale affare;
>
> e dicemi esto motto per usanza:
> 'Or ecco leggiadria di gentil core,
> per una sì selvaggia dilettanza

> lasciar le donne e lor gaia sembianza!'
> Allor, temendo non che senta Amore,
> prendo vergogna, onde mi ven pesanza.

But these, delightful as they are, *are* a mere parenthesis. More central to Dante's activity as a poet-philosopher of love are the poems of Cavalcanti 'maggiore', of Cavalcanti the tragedian. And it was by way of reaction to Cavalcanti the tragedian that, taking his cue from Guinizzelli, Dante inaugurated his 'praise style' or 'stilo de la loda'. The circumstances are recorded in the *Vita nuova*. Cross-questioned by a group of Beatrice's companions as to the object of his love, Dante replies that it lies henceforth 'in those words praising my lady', 'in quelle parole che lodano la donna mia'. The affirmation is decisive, for it is at this point that Dante comes into his own as a *stilnovista*. Having learnt from Cavalcanti the virtues of tact and economy, he now discovers in the substance and psychology of the hymn a nucleus of ideas congenial to him. With the obsessive self-preoccupation of Cavalcanti at last set aside ('mi parea di me assai avere manifestato', he confesses in *Vita nuova* XVII), he is free to focus his attention in a disinterested fashion on the goodness and beauty of his lady, aspiring as he does so to a new spiritual consciousness—to a consciousness at once purer and, in the sense of being open now to the distinctively human possibility of self-transcendence through understanding, more properly progressive.

Dante's cycle of praise poems begins with *Donne ch'avete intelletto d'amore*, a solemn canzone of five fourteen-line stanzas in hendecasyllables.[18] Conscious of the elevated nature of his theme, Dante begins in a confident and assertive manner: 'i' vo' con voi de la mia donna dire', line 2; 'Io dico che pensando', line 5; 'ma tratterò del suo stato gentile', line 11. His purpose, he says, is to give voice to a new and pressing state of mind ('ma ragionar per isfogar la mente', line 4), though one scarcely susceptible to complete statement in words ('non perch'io creda sua laude finire', line 3). What follows—Dante's celebration of Beatrice's goodness and beauty as a donna-angelo—is familiar enough. The key emphases are all well documented. But what matters about them is their ordering, Dante's promotion of the visionary and ontological over the moral and aesthetic. First, then, comes the idea of Beatrice in glory, of Beatrice as one of the elect:

> Angelo clama in divino intelletto
> e dice: 'Sire, nel mondo si vede
> maraviglia ne l'atto che procede
> d'un'anima che 'nfin qua su risplende.'
> Lo cielo, che non have altro difetto
> che d'aver lei, al suo segnor la chiede,
> e ciascun santo ne grida merzede.
>
> (lines 15–21)

Only subsequently does Dante turn to the moral and the aesthetic, to the celebration of his lady's goodness ('or voi di sua virtù farvi savere', line 30) and beauty ('Dice di lei Amor: "Cosa mortale | come esser pò sì adorna e sì pura?" ', lines 43–4)— these, however, enjoying a fresh consecration, a new and properly (as distinct from metaphorically) transcendent substance. No longer is it a matter of similitude and hyperbole. Instead, Beatrice is properly and univocally, and indeed on analogy with Christ, a special presence, a means of grace, a new Messiah even, within the economy of God's plan. The analogy, it is true, has yet to be worked out fully, something Dante will do in the *Vita nuova*; but clear already is his shifting of Guinizzelli's tentative likenesses ('Tenne d'angel sembianza') on to the plane of ontological consistency. And this is the consistency flowing into and informing the twin peaks of Dante's new 'praise style', the two sonnets *Tanto gentile e tanto onesta pare* and *Vede perfettamente onne salute*, preceded in the *Vita nuova* by the disquisitional sonnet *Amore e 'l cor gentil sono una cosa*. *Amore e 'l cor gentil* is a tribute to Guinizzelli, the 'saggio' of line 2. Like the Sicilian *tenzoni* or debate poems of Giacomo da Lentini and company, it sets out to state what love is and how it comes about, underlining as it does so the importance of *Al cor gentil* for Dante's new 'intelletto d'amore'. Love, he says, is a disposition of the noble heart activated by a desire for union with the object of sense perception. Through it, the lover stands to be recreated morally and intellectually, to be confirmed in the dignity of his (or her) properly human nature. But it is only in *Tanto gentile* and *Vede perfettamente* that the triumph of the Dantean *stil novo* emerges fully.[19] *Tanto gentile* is a hymn to

Beatrice, but like all hymns it looks both outwards and inwards, outwards to the goodness and beauty present in madonna, and inwards to the state of spiritual 'complacency' or sweet acquiescence generated by the act of perception; hence the poem's oscillation between the objective and the subjective, between the good *other* than self and the properly experiential:

> Tanto gentile e tanto onesta pare
> la donna mia quand'ella altrui saluta,
> ch'ogne lingua deven tremando muta,
> e li occhi no l'ardiscon di guardare.
>
> Ella si va, sentendosi laudare,
> benignamente d'umiltà vestuta;
> e par che sia una cosa venuta
> da cielo in terra a miracol mostrare.
>
> Mostrasi sì piacente a chi la mira,
> che dà per li occhi una dolcezza al core,
> che 'ntender no la può chi no la prova:
> e par che de la sua labbia si mova
> un spirito soave pien d'amore,
> che va dicendo a l'anima: 'Sospira.'

The sonnet is rich in all the devices of the earlier love poems, in their prolepsis (inversion), *replicatio* (repetition), *conversiones*, and hyperbole (overstatement). But now there is a sense of form, not as imposition or as addition, but as the *profile* of experience, as that whereby experience actually *is* intelligibly; so, for example, the initial prolepsis and repetition ('Tanto gentile e tanto onesta pare | la donna mia'), confirming at once both the object and the quality of Dante's contemplation, the splendour of *what* he sees and the regularity and insistence of *how* he sees; so too the *conversiones* of lines 6 and 10 ('umiltà' and 'dolcezza') and the anaphora of lines 7 and 12 ('par che . . . par che . . .'), suggesting a unique firmness of intuition and a sustained attitude of wonder, of rapt meditation; so too the delicately expressive hyperbole or *superlatio* of the whole, which, culminating as it does in the final *emphasis* of 'Sospira', stresses the poem's confessional intensity, its fervour in the act of praise. But—and this is the point—none of this is 'added on'. Nothing is decorative. What *is*, is in and through the word, the rhythm of the word being the rhythm of meditation, the sweetness of the word the sweetness of perception, the fluidity of the word the fluidity of thought and feeling. The two planes, the formal and experiential, coalesce. They become one in the irreducible spiritual-linguistic unity of the poem as a whole.

As a further example we may take *Tanto gentile*'s companion-piece *Vede perfettamente onne salute*. Here too Dante's theme is exemplary beauty *beyond* matched by rapt submission *within*. And here too everything—sound, syntax, and, such as it is, rhetorical strategy—combines to transcribe a state of perfect spiritual intuition: hence the initial epanalepsis ('Vede . . vede' at the beginning and end of the first period), the conversions of the second quatrain ('seco vestute | di gentilezza, d'amore e di fede'), the gentle concatenation of the poem's constituent clauses ('E . . . che nulla . . . anzi . . . e . . . ma . . . Ed è . . . che . . . che'), the sweet insistence of its leading sounds, of its *d*s and *v*s and, towards the end, of its *s*s, and (again) its taking up of the whole in the final, emphatic 'che non sospiri in dolcezza d'amore', all serving to confirm, as in *Tanto gentile*, the union *in re* of form and substance, of means and message:

> Vede perfettamente onne salute
> chi la mia donna tra le donne vede;
> quelle che vanno con lei son tenute
> di bella grazia a Dio render merzede.
>
> E sua bieltate è di tanta vertute,
> che nulla invidia a l'altre ne procede,
> anzi le face andar seco vestute
> di gentilezza, d'amore e di fede.
>
> La vista sua fa onne cosa umile;
> e non fa sola sé parer piacente,
> ma ciascuna per lei riceve onore.
>
> Ed è ne li atti suoi tanto gentile,
> che nessun la si può recare a mente,
> che non sospiri in dolcezza d'amore.

Associated with these 'pure' praise poems (with which, incidentally, we might put *Ne li occhi porta la mia donna Amore* and *Di donne io vidi una gentile schiera*, neither, though, advancing conceptually or technically beyond those we have seen) are those grouped by Dante around the events of the *Vita nuova*: the death of Beatrice's father, Beatrice's death, the incident of the 'donna gentile', and Dante's thoughts of Beatrice in glory. The two poems connected in the *Vita nuova* with the death of Beatrice's father are *Voi che portate la sembianza umile* and *Se' tu colui c'hai trattato sovente*, the question and reply elements in a *tenzone* or exchange with her companions. Like the Cavalcantian poems these are poems of anxiety, but of anxiety presupposing, and thus mitigated by, Dante's new understanding of love as praise. The overall tone, therefore, is different. It is elegiac rather than dramatic, meditative rather than desperate. Here too, however, technique is called on as the principle of spiritual intelligence. In *Voi che portate* it is syntax that establishes the rhythm of sadness ('onde venite . . . Vedeste voi . . . Ditelmi, donne . . . piacciavi di restar . . . nol mi celate . . .'), while in *Se' tu colui* it is lexis and euphony ('piangi', 'pianger', 'piangere', 'pianto', 'piangendo'; 'pietà', 'dolorosa', 'triste'). Form is everywhere impressive but nowhere otiose, nowhere cumbersome. At each stage it draws the mind into an act of understanding, into a state of intelligent sympathy and compassion. Compare in this respect, not least as an indication of why some poems were included in the *Vita nuova* and others not, the two excluded poems *Onde venite voi così pensose* and *Voi, donne, che pietoso atto mostrate* with the *Vita nuova* sonnet *Voi che portate*: on the one hand, conversion ('ch'i' ho dottanza . . .'), Sicilianism ('sarebbe quella ch'è nel mio cor penta?'), and an uncertain focus of attention (what, fundamentally, are the poems about: Beatrice's or Dante's suffering?); while on the other hand, a masterpiece of tact, spare in every kind of rhetorical device, and thematically centred. First, *Voi, donne*:

> Voi, donne, che pietoso atto mostrate,
> chi è esta donna che giace sì venta?
> sarebbe quella ch'è nel mio cor penta?
> Deh, s'ella è dessa, più non mel celate.
>
> Ben ha le sue sembianze sì cambiate,
> e la figura sua mi par sì spenta,
> ch'al mio parere ella non rappresenta,
> quella che fa parer l'altre beate.
>
> 'Se nostra donna conoscer non pòi,
> ch'è sì conquisa, non mi par gran fatto,

però che quel medesmo avvenne a noi.
Ma se tu mirerai il gentil atto
de li occhi suoi, conosceraila poi:
non pianger più, tu se' già tutto sfatto.'

and then *Voi che portate*:

Voi che portate la sembianza umile,
con li occhi bassi, mostrando dolore,
onde venite che 'l vostro colore
par divenuto de pietà simile?

Vedeste voi nostra donna gentile
bagnar nel viso suo di pianto Amore?
Ditelmi, donne, che 'l mi dice il core,
perch'io vi veggio andar sanz'atto vile.

E se venite da tanta pietate,
piacciavi di restar qui meco alquanto,
e qual che sia di lei, nol mi celate.

Io veggio li occhi vostri c'hanno pianto,
e veggiovi tornar sì sfigurate,
che 'l cor mi triema di vederne tanto.

On the one hand, traditionalism, thematic confusion, and, in the tercets especially, a certain clumsiness of expression; while on the other, a poignant exploration of sorrow and a wellnigh ascetic disavowal of rhetorical strategy.

Of the poems connected with Beatrice's death, the most important is the canzone *Li occhi dolenti*, a companion-piece to *Donne ch'avete* and balanced against it in the overall structure of the *Vita nuova*. Preceding it are (in the *Vita nuova*) the canzone *Donna pietosa e di novella etate*, a curiously apocalyptic account of Beatrice's assumption as imagined by Dante in a state of fevered delirium, and (tonally wholly at odds with *Donna pietosa* and thus excluded from the *Vita nuova*) the delightfully turned *Un dì si venne a me Malinconia*; neither, however, anticipates the substance and intensity of *Li occhi dolenti*, which, like *Donne ch'avete*, inaugurates in the *Vita nuova* a 'nuova matera'—the praise of Beatrice in death—and which, again like its predecessor, gives voice to a pressing movement of the spirit ('s'i' voglio sfogar lo dolore', line 4). The keynote is indeed *dolore*, but this, following on the certainties of *Donne ch'avete*, is a sorrow which issues, not in anxiety, but in a subdued contemplation of

Beatrice in glory and, as far as the poet himself is concerned, in a kind of melancholy self-searching. Only towards the end, in the fifth stanza, does Dante give way to grief, to guilt, and to the wretchedness of isolation, even this, however, containing within itself the prospect of spiritual calm to come. In the meantime technique once more gives shape to the quality and rhythm of sadness, the rich sequence of cognate vowel sounds ('Ora, s'i' voglio sfogar lo dolore, | che a poco a poco a la morte mi mena'), the concentration of lexical and semantic groups ('pietà', 'pena', 'piangendo', 'doglia', 'tristizia'; 'alto', 'cielo', 'grazia', 'gloriosa'; 'benignitate', 'umilitate', 'vertute'), together with the frequent inversion and repetition of the poem's leading terms, serving to render accessible a state of mind marked by reflection and self-interrogation. Here too, therefore, form bears a heavy weight of responsibility, a deep answerability to the 'dittar dentro' of the spirit.

Other poems in this group include *Venite a intender* and *Quantunque volte*, grief poems commissioned according to the *Vita nuova* by Beatrice's brother on the occasion of her death, and the sonnet (with, uniquely, alternative first quatrains) *Era venuta ne la mente mia* (*Vn* XXXIV), all of them reminiscent of Cavalcanti in their sense of psychological drama. More important, however, are those 'praise' poems linked in the *Vita nuova* (XXXV-XXXVIII) with the episode of the 'donna gentile', *Videro li occhi miei*, *Color d'amore*, *L'amaro lagrimar*, and *Gentil pensero*. The interpretation of these poems is complicated by what, implicitly, Dante has to say about them in the *Convivio*; for according to the *Convivio* they are not so much erotic as esoteric in inspiration. They have to do with Dante's celebration, not of a woman, but of philosophy, and have therefore to be read allegorically. The difficulty about this is that while there is nothing in the poems to preclude such a reading, there is, equally, nothing in them to prompt it, nothing crying out for such drastic reinterpretation. With other poems it is different. *Amor, che ne la mente mi ragiona*, for instance, is similarly subject to reinterpretation in the *Convivio*; but in that case—in the case, that is, of a poem very definitely representing a further stage in Dante's development as a *stilnovista*—there is, as we shall see, much in the text to suggest an *originally* philosophical conception. But that is not the case with the 'donna gentile' poems of the *Vita nuova*, which make good sense as they stand. Taking them, then, as they come, the most important thing to notice is their almost complete technical austerity. Everything is spare. The images are spare, the syntax is spare, the flowers of rhetoric are spare, and even the sound is spare:

> Color d'amore e di pietà sembianti
> non preser mai così mirabilmente
> viso di donna, per veder sovente
> occhi gentili o dolorosi pianti,
>
> come lo vostro, qualora davanti
> vedetevi la mia labbia dolente;
> sì che per voi mi ven cosa a la mente,
> ch'io temo forte non lo cor si schianti.
>
> Eo non posso tener li occhi distrutti
> che non reguardin voi spesse fiate,
> per desiderio di pianger ch'elli hanno:
>
> e voi crescete sì lor volontate,
> che de la voglia si consuman tutti;
> ma lagrimar dinanzi a voi non sanno.

This, moreover, is a state of affairs perpetuated into the final group of *Vita nuova* praise poems, those linked in the book with Dante's return to Beatrice after the 'donna gentile' episode. The repentance sonnet *Lasso, per forza* proceeds gently to unfold a state of mind marked by misgiving and by renewed commitment, while *Deh peregrini che pensosi andate* explores through the image of the pilgrim the idea of universal sorrow at Beatrice's death. Both, however, have about them the same exiguousness of form, the same economy of means. It is only in the last sonnet of the *Vita nuova, Oltre la spera che più larga gira*, that we witness any renewed sense of 'speciousness'—in the medieval sense of 'becomingness'[20]—as spirituality, of strategy as intrinsically functional. Thematically, the sonnet marks a point of arrival. Beatrice is in glory, and the lover is finally at peace in his understanding of love as renewal through contemplation. But *Oltre la spera* is a triumph above all of form, its leading notion of peace through understanding being confirmed by the most exquisite technical symmetry—by its aural consistency ('spera', 'sospiro', 'splendore', 'spirito', 'sì parla', 'spesso', 'sì ch'io lo 'ntendo'), by its studied reiteration of terms ('vede . . . Vedela';

'parla . . . parla'), and by the neat 'nesting' of its principal emphases ('io no lo intendo . . . sì parla sottile . . . lo fa parlare . . . sì ch'io lo 'ntendo ben'):

> Oltre la spera che più larga gira
> passa 'l sospiro ch'esce del mio core:
> intelligenza nova, che l'Amore
> piangendo mette in lui, pur su la tira.
>
> Quand'elli è giunto là dove disira,
> vede una donna, che riceve onore,
> e luce sì, che per lo suo splendore
> lo peregrino spirito la mira.
>
> Vedela tal, che quando 'l mi ridice,
> io no lo intendo, sì parla sottile
> al cor dolente, che lo fa parlare.
>
> So io che parla di quella gentile,
> però che spesso ricorda Beatrice,
> sì ch'io lo 'ntendo ben, donne mie care.

With *Oltre la spera*, then, we reach a significant moment of upturn, the reinstatement of rhetoric now internally justified. Subject in the course of the experience represented by the early *rime* as a whole to the most stringent purification, it emerges now refreshed and revitalized. Far from extraneous, it is rooted henceforth in the movement of thought and feeling. This, therefore, together with *Tanto gentile* and *Vede perfettamente*, is the high point of Dante's first phase as a lyric poet, the moment of triumph preparatory to a fresh order of experimentation in the moral and allegorical *rime* to come.

(iii)

The Detto d'amore *and the* Fiore

Dante's *rime* up to the *Vita nuova*, then, witness to two things: to a gradual resolution of the concepts and themes of *fin'amor* in the idea of love as a principle of moral regeneration, and to an ever greater tact in the handling of traditional rhetorical devices. Both these tendencies, as we shall see, are confirmed in the *Vita nuova*. First, though, we must look at the so-called *Detto d'amore* and *Fiore*, belonging in a probable chronology of Dante's early work to about 1286 or 1287. Each poem survives in a single manuscript, the *Detto d'amore* in Ashburnham 1234 of the Biblioteca Medicea Laurenziana in Florence, and the *Fiore* in H 438 of the University Library (Medicine) of Montpellier, though it was clear to the first editor of the *Detto* (Morpurgo, 1888) that these were parts of one and the same manuscript belonging to the early years of the Trecento and subsequently split up. However we interpret them (and interpretation, after attribution, is the main problem with these poems),[21] they each represent a coming to terms on Dante's part with the single most important vernacular *tractatus de amore* of the late Middle Ages, the *Roman de la rose*, and this in two stages: a Guittonian-cum-Guinizzellian stage—the *Detto*—and a more dramatic and in certain respects more Cavalcantian stage—the *Fiore*.

The *Roman de la rose* is a lengthy romance of some 22,000 lines composed between about 1230 and 1280 by Guillaume de Lorris (d. *c*.1237), responsible for the first 4000 lines, and Jean de Meun (d. 1305), author of the rest. The two parts are very different. The first part by Guillaume de Lorris looks on the face of it— 'on the face it', for even here there is room for debate—to be an elegant courtly fable. The lover, falling into a dream, is introduced by Idleness (Oiseuse), gatekeeper to a genteel beau by the name of Mirth (Deduiz), into a *hortus deliciarum* or garden of delights rigorously exclusive of the sins and griefs of the world (Hatred, Greed, Envy, Sorrow, Age, Poverty, etc.) but populated by the likes of Courtesy, Gladness, and Sweetlooks, all taking their ease within. Eventually, the lover is introduced to the God of Love and initiated by him into love's sweet mysteries. Love, Amors says, brings many a pain and perplexity, but patience and perseverance are richly rewarded, whereupon the lover, mightily encouraged by what he has heard, sets off in search of the rose. Cruelly thwarted, however, by Danger, Evil Tongue, Shame, and Fear, he is comforted at last by Reason, who attempts to persuade him of the folly of his ways. Do not put your trust, she says, in enterprises of this kind, for fortune treats harshly those who tempt her. And as for the God of Love, he is better abjured altogether, for all he has to offer is disappointment and misery. But the lover is not impressed, and, with the help of Pity and Franchise, manages to get back into the enclosure and steal a kiss of the flower, at which point Evil Tongue and Jealousy, catching wind of what is going on, chase him off yet again and imprison Bel Acueil (Fair Welcome) in a specially constructed, heavily fortified tower.

With this, Jean de Meun takes over and, in a long discourse of eminently Boethian inspiration (Boethius was one of Jean de Meun's favourite authors), has Reason attempt once again to persuade the lover of his folly. She does not wish him, she says, to eschew love altogether, but simply to love more wisely. She even offers herself, in place of Fair Welcome, as his mistress. But he remains stubborn, preferring the company of Friend and of False Seeming, who between them promise to help him out by means of stealth and deception. Finally, in what amounts to a nicely judged piece of medieval romance pornography, he is able to enjoy the rose, but not before he has listened to the two great discourses of Nature and Genius urging upon him and upon mankind generally a more liberal ploughing and scattering of the seed, this being the way of immortality.

This, needless to say, is a mere sketch, and as such does no justice to the complex background, to the wealth of learning, and above all to the brilliant sense of paradox at work throughout. But it is enough even so to indicate the problems posed by the *Rose* for the reader or commentator anxious to make sense of it. There is, for example, the relationship of the poem's two parts. Is it a matter merely of iconoclasm on the part of Jean de Meun, of a crude despoiling of Guillaume de Lorris's neat refinements, or did he discern in Guillaume de Lorris an incipient meditation on love's immanent contradictions and moral perils? No one seems quite sure. And what about Jean de Meun? Is he an out-and-out realist *à la* Rustico di Filippo or Cecco Angiolieri, a professional sender-up of hallowed themes, or is he simply a university wit, good at intellectual games but not much committed to anything in particular? Again, it is difficult to say, and classical *Rose* criticism has veered now in the direction of those who see in the poem nothing more than a fit of pre-Rabelaisian irreverence (Langlois, Paré, Louis, etc.) and now in the direction of those who regard it as a comic but none the less seriously intentioned essay in affective philosophy along Boethian and Chartrian lines (Fleming, after Robertson, especially).[22]

Dante's first encounter with the *Roman*, the *Detto d'amore*, represents an oddly transitional moment between Guittonianism and the *stil novo*. Virtuoso in style—the poem is unfolded in rhymed seven-syllable couplets fraught with every kind of technical difficulty—it also has about it a nicely Sicilian-cum-Guinizzellian sense of imaginative delicacy and *ariosità*. So, for example,—singing and dancing apart (a bit popular for the *stil novo* proper)—lines 233-45, reminiscent of Giacomo da Lentini's *Madonna ha 'n sé vertute con valore* or of Guinizzelli's *Io voglio del ver la mia donna laudare*:

> E quando va per via,
> Ciascun di lei à 'nvia
> Per l'andatura gente;
> E quando parla a gente,
> Sì umilmente parla
> Che boce d'agnol par là.
> Il su' danzar e 'l canto
> Val vie più ad incanto
> Che di nulla serena,
> Ché. ll'aria fa serena:
> Quando la boce lieva,
> Ogne nuvol si lieva
> E l'aria riman chiara.

But what matters about the *Detto*, reminiscences apart, is its stress, not on the narrative aspects of Guillaume de Lorris's *Rose*, but on its psychological aspects, on *fin'amor* as a form of awareness. True, the story is there, lurking in the background; but typically the narrative is resolved in the contemplative, in the most urgent kind of emotional and intellectual self-searching. So, for example, the pleasure in service motif of the opening lines (1-16):

> Amor sì vuole, e par-li,
> Ch'i' 'n ogni guisa parli
> E ched i' faccia un detto,
> Che sia per tutto detto,
> Ch'i' l'aggia ben servito.
> Po' ch'e' m'ebbe 'nservito
> E ch'i' gli feci omaggio,
> I' l'ò tenuto o. maggio
> E terrò giamà' sempre;
> E questo fin asempr'è
> A ciascun amoroso,

> Sì ch'Amor amoroso
> No.gli sia nella fine,
> Anzi ch'e' metta a fine
> Ciò ch'e' disira avere,
> Che val me' ch'altro avere.

or the adoration and thankfulness motif of lines 29–37:

> E dònati in presente,
> Sanz'esservi presente,
> Di fino argento o d'oro
> Perch'i' a.llui m'adoro
> Come leal amante.
> A.llu' fo grazze, amante
> Quella che d'ogne bene
> È sì guernita bene
> Che'n le' non truov'uon pare . . .

or the hope and trust motif of lines 63–73:

> E chi la porta in grado
> Il mette in alto grado
> Di ciò ched e' disia:
> Per me cotal dì sia!
> Per ch'i' già non dispero,
> Ma ciaschedun dì spero
> Merzé, po' 'n su' travaglio
> I' son sanza travaglio,
> E sonvi sì legato
> Ch'i' non vo' che legato
> Giamai me ne prosciolga . . .

At each stage it is the condition itself of knowing, loving, and rejoicing that engages the poet's interest. Each episode sees the lover further immersed in the moral and psychological complexities of his new profession. And this in turn justifies the poem's relentlessness on the technical front; on the one hand a claustrophobic sense of love as obsession, as fixation, as dogma, as captivity to a creed; while on the other the ceaseless reiteration of rhyme, sound, and syntax. In both senses, the *Detto* is an essay in constraint and in circularity, with form—in its recognizably Guittonian guise (or rather Brunettian guise, for not far away stands the *Tesoretto* of Brunetto Latini)— bearing as always in Dante a heavy expressive responsibility.

More formidable, however, as an encounter with the *Rose* is the *Fiore*, a cycle of 232 sonnets reflecting again, though less assiduously, the poetic *moeurs* of Guittonian and Tuscan lyricism. Here, Dante's reception of the *Rose* is more ample; for where the *Detto* had been fashioned on the basis of Guillaume de Lorris's *Rose*, the *Fiore*, though in a freely restructured and immeasurably more economic form, encompasses the whole work. And this now is the point, for it is the *Fiore*'s more considered economy that complicates its interpretation, that invites us to see in it, over and above acquiescence in a sparkling display of Gallic wit, a 'question of love' in some sense anticipating the *Vita nuova*.

The initial *mise-en-scène*, a masterpiece of narrative and ideological succinctness, establishes the ethos of *fin'amor*. Amore instructs Amante on the arduousness of the way. Patience, he says, is all. Whatever love's trials and tribulations, joy awaits its faithful follower:

> 'Ma chi mi serve, per certo ti dico
> Ch'a la mia grazia non può già fallire,
> E di buona speranza il mi notrico
> Insin ch'i gli fornisca su' disire.
>
> E pensa di portar in pacïenza
> La pena che per me avrà' a sofrire
> Inanzi ch'io ti doni mia sentenza;
>
> Ché molte volte ti parrà morire:
> Un'ora gioia avrai, altra, doglienza;
> Ma poi dono argomento di guerire.'
>
> (III. 11–14 and IV. 9–14)

Whereupon Amante professes loyalty and obedience:

> Con grande umilitate e pacïenza
> Promisi a Amor a sofferir sua pena,
> e ch'ogne membro, ch'i' avea, e vena
> Disposat'era a farli sua voglienza ...
>
> (V. 1–4)

and takes an oath of allegiance ('E saramento gli feci e omaggio', III. 4) sealed formally in a little ceremony of the keys ('Con una chiave d'or mi fermò il core', IV. 1).

With this the story proper once again gets under way,

Amante being all eager to pursue his pre-eminently sensuous love ('Allor mi venni forte ristrignendo | Verso del fior, che.ssì forte m'ulìo', VI. 5–6) and to embark on the strategies necessary to encompass its end. The whole sequence of events in these sonnets—Amante's first repulse by Schifo, the intercession of Franchezza, of Pietà, and of Venus, and Amante's second repulse by Gelosia—is in the nature of a burlesque, the comic irony of which is only intensified by the sage intervention of Ragione (IX), who advises the lover of the folly of his ways:

> 'I' credo che.ttu à' troppo pensato
> A que' che.tti farà gittar in vano,
>
> Ciò è Amor, a cui dat'ài fidanza.
> Ma.sse m'avessi avuto al tu' consiglio,
> Tu non saresti gito co.llui a danza:
>
> Ché, sie certano, a cu' e' dà di piglio,
> Egli 'l tiene in tormento e malenanza,
> Sì che su' viso nonn-è mai vermiglio.'
>
> (IX. 7–14)

But Reason's advice to the effect that the lover should withdraw from the dance is peremptorily dismissed (X. 13–14), and the process of self-ingratiation prescribed by Andreas Capellanus in the *De amore* and culminating in the kiss duly inaugurated. Crossing himself reverently, the lover takes his pleasure:

> Sì ch'i' alor feci croce de le braccia,
> E sì 'l basciai con molto gran tremore . . .
>
> (XX. 12–13)

But his happiness is fragile. Gelosia is aroused by the gossip Malabocca, and Amante is put forth rudely from the garden:

> Del molto olor ch'al cor m'entrò basciando
> Quel prezïoso fior, che tanto aulia,
> Contar né dir per me non si poria;
> Ma dirò come 'l mar s'andò turbando
>
> Per Mala-Bocca, quel ladro normando,
> Che se n'avide e svegliò Gelosia
> E Castità, che ciascuna dormia;
> Per ch'i' fu' del giardin rimesso in bando.
>
> (XXI. 1–8)

Here, just where, in the *Roman*, narrative responsibility passes from Guillaume de Lorris to Jean de Meun, Dante's tone undergoes perceptible intensification. The lover, victim of his own concupiscence, of his own reckless submission of reason to desire, pauses to ponder his predicament, the infernal predicament, he seems to sense, of one embarked on the way of wilful self-destruction:

> Quand'i' vidi i marosi sì 'nforzare
> Per lo vento a Provenza che ventava,
> Ch'alberi e vele e ancole fiaccava,
> E nulla mi valea il ben governare,
>
> Fra me medesmo comincià' a pensare
> Ch'era follïa se più navicava,
> Se quel maltempo prima non passava
> Che dal buon porto mi facé' alungiare:
>
> Sì ch'i' allor m'ancolai a una piaggia,
> Veggendo ch'i' non potea entrar in porto:
> La terra mi parea molto salvaggia.
>
> I' vi vernai co.molto disconforto.
> Non sa che mal si sia chi non asaggia
> Di quel d'Amor, ond'i' fu' quasi morto.
>
> Pianto, sospiri, pensieri e afrizione
> Ebbi vernando in quel salvaggio loco,
> Ché pena de.ninferno è riso e gioco
> Ver' quella ch'i' soffersi a la stagione
>
> Ch'Amor mi mise a tal distruzïone
> Ch'e' no.mi die' soggiorno assà' né poco:
> Un'or mi tenne in ghiaccio, un'altra 'n foco.
>
> (XXXIII; XXXIV. 1–7)

Fortunately, though, the mood does not last. Confronted again by Reason and her injunction to found his happiness on something more secure than fortune, he falls into his old ways. What is all this, he says, about renouncing love? Reason, he insists, in counselling him thus, is a scoundrel and a sophist. By inviting him not to love at all (for this, as he understands it, is the upshot of what she is saying) she is inviting him both to perjure himself and to forsake the business of procreation, something which as an honest man he will never do. Notwithstanding, therefore, Reason's careful account of what she means—it is, she explains, a question, not of rejecting love, but of loving in an orderly way (XLV)—he commits himself come what may to conquering the rose:

> Quando Ragion fu assà' dibattuta
> E ch'ella fece capo al su' sermone,
> I' sì.lle dissi: 'Donna, tua lezione
> Sie certa ch'ella m'è poco valuta,
>
> Perciò ch'i' no.ll'ò punto ritenuta,
> Ché no.mi piace per nulla cagione;
> Ma, cui piacesse, tal amonizione
> Sì gli sarebbe ben per me renduta.
>
> Chéd i' so la lezion tratutta a mente
> Pe.repètall'a gente cu' piacesse,
> Ma già per me nonn-è savia nïente:
>
> Ché fermo son, se morir ne dovesse,
> D'amar il fior, e 'l me' cor vi s'asente,
> O 'n altro danno ch'avenir potesse.'
>
> (XLVI)

Reason, then, is no companion for the sage lover. Her sole purpose is to stand in the way of his rightful happiness. More helpful is Amico, a true friend in need, and scornful from the very beginning of Reason's advice:

> 'Guarda che non sie accettato
> Il consiglio Ragion, ma da te il buglia,
> Ché' fin' amanti tuttor gli tribuglia
> Con quel sermon di che.tt'à sermonato.'
>
> (XLIX. 5–8)

Love for him is a matter of self-indulgence through strategy and deception. Unfolding his *intelletto d'amore* with massive consistency and deep cynicism, he describes for Amante's benefit the importance for lovers of flattery, intrigue, versatility, adaptability, and, above all, of sustained dissimulation. 'Far sembiante' is a typical formula: 'Di pianger vo' che faccie gran sembianti' (LIII. 7); 'Sembianti fa che sie forte crucciato' (LV. 7); 'fa sembiante che nonn-ài che farne' (LV. 11); 'Non far sembiante d'averla veduta' (LXII. 10), and Amante, neatly, is a little shocked. May God keep him, he says (LXVIII. 3–8), from actually wishing harm to others. Needs must, though, and in the end he takes on board his friend's advice:

> 'Po' mi convien ovrar di tradigione
> E a.tte pare, Amico, ch'i' la faccia,
> I' la farò, come ch'ella mi spiaccia,
> Per venir al di su di quel cagnone' [i.e. Malabocca].
>
> (LXX. 1–4)

Falsembiante's discourse, Amico's writ large, is a study in the ways and means of fraud generally. True, there is about it an element of satire, a pointing-of-the-finger at one group in particular of offenders (clerics), but that is by the by. The basic problem—and in this sense Falsembiante's intervention represents a standing-back from the narrative proper, a kind of meditative interlude—is that of form in relation to substance. It is a question of how far, in human activity, words do or not match up with deeds. The problem is characteristically Dantean. Take, for example, this passage from the sonnet (to Cino) *Io mi credea del tutto esser partito* (lines 12–14):

> Però, se leggier cor così vi volve,
> priego che con vertù il correggiate,
> sì che s'accordi i fatti a' dolci detti

or these lines from *Inferno* IV (147) and XXXII (12):

> che molte volte al fatto il dir vien meno

> sì che dal fatto il dir non sia diverso

and compare them with Falsembiante in CIII. 9–11:

> 'Così vo io mutando e suono e verso
> E dicendo parole umili e piane,
> Ma molt'è il fatto mio a.dir diverso . . .'

The point is that love, or rather the professing of love, raises the problem of form and content. Amore's discourse in sonnets I–V is full of noble thoughts: thoughts of honour, fair play, patience, self-denial, service, and humility. It is morally unexceptionable. But nobility as professed by Falsembiante and by Amico is nothing but the means to an end. It is empty, systematically disingenuous, bent solely on profit; hence the unfolding of his discourse in terms of fraud ('re de' barattier'',

LXXXVII. 6; 14; CXXVII. 4, etc.), of hypocrisy, and of thievery. He is, he says, a disciple of the Antichrist, of the Arch-Deceiver:

> 'I' sì son de' valletti d'Antecristo,
> Di quel' ladron' che dice la Scrittura
> Che fanno molto santa portatura,
> E ciaschedun di loro è ipocristo.'
>
> (CXXIII. 1–4)

But if Falsembiante's is an account of love as deceitfulness, then the Vecchia's is an account of love as venality. Type of the old-fashioned hypocritical *entremetteuse*, her only purpose is to make a pile, or rather (for she personally has had her day) to see her young ward Bellacoglienza make a pile. Proceeding, then, by way of the most cloying sentimentality—a sentimentality, however, scarcely concealing her rapaciousness—she advises Bellacoglienza to accept without ado Amante's material advances ('gran peccato faresti | Se 'l su' presente tu gli rifusassi', CXLII. 8–9) and continues thus to outline her own time-honoured *ars amatoria*:

> 'E s'alcun altro nonn-à che donare,
> Ma vorràssi passar per saramenta,
> E dirà che.lla 'ndoman più di trenta
> O livre o soldi le dovrà recare,
>
> Le saramenta lor non dé pregiare,
> Chéd e' non è nessun che non ti menta
> . . . . . .
> Per che già femina non dee servire
> Insin ch'ella non è prima pagata . . .'
>
> (CLXXIX. 1–6 and 10–11)

—an attitude all of a piece with that of Amico, also inclined to see love as cashing in ('E se.lla donna prende tu' presente, | Buon incomincio avrà' di far mercato', LV. 1–2). Women everywhere, the Vecchia insists, have but one object, to rook the customer whoever he is—count, cavalier, citizen, or whoever—for all he has got; for that, she says, is what love is all about.

Everyone, then,—Amante, Amico, Falsembiante, the Vecchia and her whole (erstwhile) clientele—joins the dance for reasons of self-interest. Love's devaluation is complete.

Emptied of its nobler connotations, the term 'amore' is understood now to mean simply extortion, milking the victim for every last penny: 'E saccie far sì che ciascuno adenti', proclaims the Vecchia, 'Insin ch'a povertà gli metterai' (CLX. 7–8).

Each of the leading characters, Ragione, Amico, Falsembiante, and the Vecchia, has now had his say, and it remains only to trace the dramatic denouement of the tale. Here as before it is a matter of burlesque, of an extravagant account, mock-epic in tone, of the triumph of *amor concupiscentiae*. Solemnly, Amante renews his oath to Amore on the instruments of amorous assault:

> Per far le saramenta sì aportaro,
> En luogo di relique e di messale,
> Brandoni e archi e saette; sì giuraro
> Di suso, e disser ch'altrettanto vale.
>
> (CCXIX. 9–12)

and, in what amounts now to flagrant parody (cf. Mark 6: 8; Luke 22: 35), sets out, staff and purse in hand, on his final pilgrimage:

> Ch'un bordon noderuto v'aportai,
> E la scarsella non dimenticai,
> La qual v'apiccò buon mastro divino.
>
> (CCXXVIII. 6–8)

The ineptly managed rape is followed by a little *Te Deum*, and by copious acknowledgement on the part of the lover of all those who have made this possible for him, especially Bellacoglienza and his good friend Amico. Reason, though, is excluded. Her malevolence, quite rightly, has availed her naught:

> Ma di Ragion non ebbi sovenenza,
> Che.lle mie gioie mi credette aver tolte.
> Ma contra lei i' ebbi provedenza,
> Sì ch'i' l'ò tutte quante avute e colte.
>
> (CCXXXI. 11–14)

What, then, are we to make of it all?

THE *DETTO D'AMORE* AND THE *FIORE*

There are three possibilities. The first is to see the *Fiore* as an exercise in bourgeois 'realism', urban, comic, and reductionist.[23] Looked at in this way—in a way, that is, reflecting the generally (though not universally) accepted view of the *Roman*—it fills out the technical history of Dante's other essays in the 'comic' style, of the Forese *tenzone* and of certain episodes in the *Inferno*. The problem with this view is that it is itself reductionist. It takes no account of the essentially 'oppositional' structure of the *Fiore*, of its recourse to antithesis as the means of argumentation. Consider, for example, the following passage from Amore's initial discourse to the lover in sonnets I-V:

> 'Ed ogn'altra credenza metti a parte,
> Né non creder né Luca né Matteo
> Né Marco né Giovanni.'
>
> (v. 12–14)

At once the issue is plain. One way of loving—the 'courtly-amorous' way—is being set off against another—the Christian-biblical way—with Amore's commandments to Amante caricaturing those of Christ and of the prophets: 'Fa che m'adori, ched i' son tu' deo' (v. 11; cf. Exod. 20: 3: 'you shall have no other gods before me'); 'E solo a lui servir la mia credenza | È ferma, né di ciò mai nonn-alena' (v. 5–6; cf. Mark 12:30: 'Love the Lord your God with all your heart, with all your soul, with all your mind, and with all your strength. This is the first commandment'); 'I' son segnor assà' forte a servire . . .' (III. 10–11; cf. Matt. 16:24: 'If anyone would come after me, let him deny himself, take up his cross, and follow me'). And if to this we add the economy of the poem, its elimination of everything marginal to the moral problem (Nature and Genius, with their notion of love as survival of the species, are cases in point), then what we have is a discourse designed, not simply to entertain, but to alert the reader to a matter ripe for resolution.

This, therefore, is the second possibility: to see the *Fiore*, not as a celebration, but as a critique of *fin'amor*, as a work designed to initiate the reader in an act of understanding. And this, as we have just noted, is a view with much to recommend it. So succinct is the poem, so clever in its antitheses, that it seems to breathe dialectic. Here too, though, there is a difficulty, arising now, not so much from the poem itself, as from its relationship with the *Rose*. If the *Rose* is a critique of *fin'amor* and Dante is following suit, fine. If instead the *Rose* is an essay in bourgeois realism, and Dante, having seen and appreciated this, is none the less choosing to make an issue of it, again fine. But how can we be sure of this? How can we be sure (*a*) of what Jean de Meun, in his massive and massively contradictory 18,000 lines, is saying, and (*b*) of Dante's independence of spirit *vis-à-vis* his text? We cannot.

But all is not lost, for the function if not the interpretation of the *Fiore* in Dante is readily defined—that function being to confirm once and for all the presence and decisiveness throughout his work of Guillaume de Lorris's and of Jean de Meun's marvellously kaleidoscopic initiative in the vernacular. Its material presence in Dante is easily demonstrated. Wherever we look, in the *Vita nuova*, in the *Convivio*, or in the *Commedia*, there it is, prompting this or that image or line of argument; so, for example, for the *Vita nuova*, the age motif (II. 1; *Roman* 21), the salutation motif (III. 1; *Roman* 2361-4), the *malparlieri* motif (X. 1; *Roman* 3493 ff. and 3550 ff.), the distance and departure motif (VII. 1-2; *Roman* 2287-301), the disguise and dissimulation motif (IV. 1 and V. 3; *Roman* 2375-8), the stream motif (IX. 4 and XIX. 1; *Roman* 1385-8), the suffering and transfiguration motif (XIV.4; *Roman* 2379-85), the dream and enigma motif (III. 15; *Roman* 1-20), and the pilgrimage motif (XL; *Roman* 21316 ff.); for the *Convivio*, the folly of wealth argument (IV. xii. 5 and xiii. 11; *Roman* 5051-60 and 5160-74), the precession of the equinoxes argument (II. xiv. 11; *Roman* 16777-92), the poor man's freedom from anxiety argument (IV. xiii. 12; *Roman* 5250-7), the unworthy descendant argument (IV. vii. 8-9; *Roman* 18725-31), the avarice of doctors and lawyers argument (III. xi. 10; *Roman* 5061-4), and the element of truth in all philosophy argument (IV. xxi. 3; *Roman* 7031-2); and for the *Commedia* the nature and art discourse in *Inferno* XI. 97-105 (*Roman* 15989-98); the progressive decline of history argument in *Inferno* XIV. 103-14 (*Roman* 20166-73); and, most conspicuous of all, the determinism and free-will discourses in *Purgatorio* XVI:

'Et se les destinees tienent
toutes les choses qui avienent,

> si con cist argumanz le preuve
> par l'apparance qu'il i treuve,
> cil qui bien euvre, ou malement,
> quant il nou peut fere autrement,
> quel gré l'an doit donc Diex savoir
> ne quel peine an doit il avoir?'

'If destiny controlled everything that happened, as going by these arguments seems to be the case, and men do well or ill because they have no choice, then what reward or punishment do they deserve from God?'

(*Roman* 17125–32)

There at every turn, ready and open-handed, is the *Roman*, equally at home in the comic and the tragic, in the sordid and the sublime. Every emphasis, every nuance, every imaginative and ideological strategy is there for Dante to contemplate and to make his own. But that is not all, for the *Roman de la rose* represented also a challenge, an invitation to do as much again, and more besides, in Italian. There it stood, larger than life, bestriding its complex tradition of Neoplatonism, of neo-Aristotelianism, of neo-Chartrianism (the mythical cosmologies of Bernardus Silvestris and Alan of Lille), of Boethian moralism, and of Christian-mystic speculation *de amicitia spirituali*. It was a masterpiece of 'translation', a model of Romance vernacular appropriation.[24] Dante could not but be impressed. He could not but be struck by the ease and accomplishment of it all, by the elegant inventiveness of Guillaume de Lorris and by the sheer omnicompetence of Jean de Meun. It is in this context, then, that we have to understand and to appreciate the *Fiore*, confirmation, were it needed, of great things afoot, of lines of communication destined to issue in the triumph, not simply of the *Vita nuova* and of the *Convivio*, but of the more distant *Commedia*.

(iv)

The Vita nuova

The *Vita nuova* is the first large-scale manifestation of what we spoke of earlier as Dante's characteristic tendency towards self-organization.[25] All his works reflect in one way or another this same tendency, this same desire to identify the 'reasons' of his being and activity so far, and, as far as may be, to resolve the many, the various, and at times the recalcitrant, in the single, the continuous, and the inherently consistent; hence their nature as works of self-interrogation, as works in which each constituent element in turn is subject to scrutiny and evaluation in the light of the whole. Hence too their underlying seriousness of purpose, and, as a guarantee of that seriousness, their search for expressive integrity. In each of these respects the *Vita nuova* is exemplary; for nowhere—the *Comedy* apart—do we witness more clearly than here Dante's concern for moral and intellectual intelligibility, for self-elucidation in and through the word.

But, again as we noted earlier, the *Vita nuova* has a sharp edge. Introspective and confessional as it is, it reaches out through confession to resolve a public issue, the long-standing and contentious question of love. The opposition is twofold. On the one hand there is the 'realist' solution represented by the *Roman de la rose* and by the *Fiore*, and on the other the Cavalcantian solution of *Donna me prega*. The opposition *Vita nuova–Roman de la rose* (*Vita nuova–Fiore*) is open and shut. The lover in the *Fiore*, following his archetype in the *Roman*, is stubborn in his rejection of Ragione and of her advice to seek satisfaction in the secure and lasting joy of the rational soul. She is, he says, a prudish hypocrite, intent on dissuading him from love altogether (XL and XLI). He will have nothing to do with her. Dante of the *Vita nuova*, by contrast, is wholly obedient to the promptings of reason, for it is through love as guided by reason that man encompasses his highest end and happiness. Never, therefore, was he tempted to disobey or to dismiss her, to strike out recklessly on his own:

> E avvegna che la sua imagine, la quale continuatamente meco stava, fosse baldanza d'Amore a segnoreggiare me, tuttavia era di sì nobilissima vertù, che nulla volta sofferse che Amore mi reggesse sanza lo fedele consiglio de la ragione in quelle cose là ove cotale consiglio fosse utile a udire.
>
> (II. 9)

The passage cannot be read other than as a corrective to the sensualism of the *Fiore*, to the unwise or foolish love (*foll'amor*) celebrated in the *Roman*.

More subtly, however, this same passage witnesses to the anti-Cavalcantian stance of the *Vita nuova*, to its wish to vindicate over and against Cavalcanti the notion of love as a principle, not of death, but of life, of spiritual renewal through understanding. The middle part of Chapter II (4–9), with its 'wonder and oppression of the spirits' motif, is, it is true, Cavalcantian enough; the prospect of joy in Beatrice's presence is accompanied by fear and trembling ('cominciò a tremare sì fortemente'), by a powerful sense of oppression (' "Ecce deus fortior me, qui veniens dominabitur michi" '), and by copious complaint ('cominciò a piangere, e piangendo disse queste parole'). But it is precisely here, in mid Cavalcanti, that Dante confirms the un-Cavalcantian association of love and reason, the mutual inherence of the affective and the cognitive in a single process of seeing, willing, and becoming.[26] The lines, then, are drawn. The *Vita nuova*, at one level intimately apologetic, assumes a public face.

With this, we are ready to retrace with Dante the steps of his literary-ideological journey in the *Vita nuova*, but not before pausing over the all-important first chapter; for here Dante indicates how the book is to be read. Imagining the events of his early years, and the texts they occasioned at the time, as constituting the 'book of his memory' ('libro de la mia memoria'), he casts himself in the role of a scribe, of an editor, and of a commentator. As a scribe he will simply copy from one manuscript to another. He will be a straightforward amanuensis. But as an editor he will be selective, choosing and copying only those items of particular significance, those most apt to yield up the inner meaning or *sententia* of the whole. Finally, as a commentator he will elucidate that meaning, pausing in the course of the narrative to emphasize its underlying rationale (Chapters XI, XXV, and XXIX, on the psychology of the *saluto*, on the integrity of form in vernacular lyric art, and on the properties of the number nine, respectively). Thus the work is 'layered' in conception, each successive intervention of editor and commentator drawing the reader into a securer possession of its leading idea.[27]

First, in Dante's ideal reconstruction of his experience so far as a vernacular poet-philosopher of love, come those poems celebrating his earliest encounter with Beatrice, and more particularly the various devices—most prominently that of the screen lady or *donna schermo*—necessary for preserving its decorum and social respectability. The tone of these chapters is ambiguous, inclined to oscillate between the 'courtly' on the one hand and the mystical and ecstatic on the other; so, for example, on the mystical and ecstatic side, the glossatorial intervention of Chapter XI, eminently Victorine in flavour[28] and, in its sense of love as disposition (as distinct from acquisition), anticipatory of developments to come:

Dico che quando ella apparia da parte alcuna, per la speranza de la mirabile salute nullo nemico mi rimanea, anzi mi giugnea una fiamma di caritate, la quale mi facea perdonare a chiunque m'avesse offeso; e chi allora m'avesse domandato di cosa alcuna, la mia risponsione sarebbe stata solamente 'Amore', con viso vestito d'umilitade. E quando ella fosse alquanto propinqua al salutare, uno spirito d'amore, distruggendo tutti li altri spiriti sensitivi, pingea fuori li deboletti spiriti del viso, e dicea loro: 'Andate a onorare la donna vostra'; ed elli si rimanea nel luogo loro. E chi avesse voluto conoscere Amore, fare lo potea mirando lo tremare de li occhi miei. E quando questa gentilissima salute salutava, non che Amore fosse tal mezzo che potesse obumbrare a me la intollerabile beatitudine, ma elli quasi per soverchio di dolcezza divenia tale, che lo mio corpo, lo quale era tutto allora sotto lo suo reggimento, molte volte si movea come cosa grave inanimata. Sì che appare manifestamente che ne le sue salute abitava la mia beatitudine, la quale molte volte passava e redundava la mia capacitade.

But what matters about these chapters is not so much their anticipation of attitudes to come as their acknowledgement and dismissal of attitudes now superseded, of ways of thinking and of writing about love rooted in the now tired practices of the Guittoniani. That, within the economy of the book as a whole, is their function, to signal a moment of thought overtaken by the new, by the ideologically and technically more refined. Ideologically, then, it is a question of love as expectation and reward (*guerdon*). Everything depends on that, on Beatrice's willingness either to oblige or not to oblige. If she obliges, fine; but if she does not, or cannot, then all is catastrophe and despair. So, for example, the following extract from the poem *Ballata, i' voi che tu ritrovi Amore* (*Vn* XII), a faithful rehearsal of the old servant–mistress motif coming down from Provence:

> Dille: 'Madonna, lo suo core è stato
> con sì fermata fede,
> che 'n voi servir l'ha 'mpronto onne pensero:
>   tosto fu vostro, e mai non s'è smagato.'
> Sed ella non ti crede,
> dì che domandi Amor, che sa lo vero:
>   ed a la fine falle umil preghero,
> lo perdonare se le fosse a noia,
> che mi comandi per messo ch'eo moia,
> e vedrassi ubidir ben servidore.
>
>   (lines 25–34)

Technically, it is a question of set pieces in the Tuscan manner, of the traditional enigma sonnet (*A ciascun'alma presa e gentil core*, III), of the *planctus* or *lamentanza* (*O voi che per la via* and *Piangete, amanti*, VII and VIII), and of the *improperium* or accusation poem (the double sonnet *Morte villana*, VIII), all unfolded, as in the case of these lines from *O voi che per la via d'Amor passate* and from *Morte villana*, with ample appeal to the strategies of the Guittoniani, to binary expressions of the type 'attendete e guardate', to apostrophe, to inversion, and to word-play of the kind 'torto tortoso':

> O voi che per la via d'Amor passate,
> attendete e guardate
> s'elli è dolore alcun, quanto 'l mio, grave;
> e prego sol ch'audir mi sofferiate,
> e poi imaginate
> s'io son d'ogni tormento ostale e chiave.
>
> Amor, non già per mia poca bontate,
> ma per sua nobiltate,
> mi pose in vita sì dolce e soave,
> ch'io mi sentia dir dietro spesse fiate:
> 'Deo, per qual dignitate
> così leggiadro questi lo core have?'
>
>   (*O voi che per la via*, lines 1–12)

> Morte villana, di pietà nemica,
> di dolor madre antica,
> giudicio incontastabile gravoso,
> poi che hai data matera al cor doglioso
> ond'io vado pensoso,

> di te blasmar la lingua s'affatica.
> E s'io di grazia ti voi far mendica,
> convenesi ch'eo dica
> lo tuo fallar d'onni torto tortoso,
> non però ch'a la gente sia nascoso,
> ma per farne cruccioso
> chi d'amor per innanzi si notrica.
>
> (*Morte villana*, lines 1–12)

Cavalcando l'altr'ier, admittedly, introduces a note of variety, of welcome irresponsibility, but these by and large are poems witnessing to Tuscanism at its most earnest. Again, however, that is their role in the *Vita nuova*. Their role is to testify to a phase of Dante's experience passionately lived out, but destined within the perspective of the whole to give way to the new, to the specifically stilnovistic.

The next phase, the Cavalcantian phase, is as we have seen characterized (*a*) by a sense of love as conflict, and (*b*) by a simplification of poetic means. This, in the *Vita nuova*, is ushered in by the intervention of Amore, who bids Dante lay aside the intrigues and strategies he has been using so far, the techniques of enigma, screen-ladies, vituperation, and lamentation, in favour of a more direct approach, of a more immediate declaration of his love for Beatrice: 'Fili mi,' Amore says in XII. 3, 'tempus est ut pretermictantur simulacra nostra' ('It is time now, my son, to lay aside our deceits'). And then, more cryptically, comes an intimation of anguish, confirmation of the pain awaiting the poet in the accomplishment of true understanding and happiness. Amore, Dante suggests, stands at the centre of a circle, he himself lingering on the edge. Love, in other words, rather like God in the theological use of this image, is at the central point of all-knowing and all-comprehending, while Dante the lover is as yet marginal and estranged, caught up in uncertainty and anxiety; hence Love's weeping and his exhortation to Dante not to ask for more than can be properly understood and assimilated:

e riguardandolo, parvemi che piangesse pietosamente, e parea che attendesse da me alcuna parola; ond'io, assicurandomi, cominciai a parlare così con esso: 'Segnore de la nobiltade, e perché piangi tu?' E quelli mi dicea queste parole: 'Ego tanquam centrum circuli, cui simili modo se habent circumferentie partes; tu autem non sic.' Allora, pensando a le sue parole, mi parea che m'avesse parlato molto oscuramente; sì ch'io mi sforzava di parlare, e diceali queste parole: 'Che è ciò, segnore, che mi parli con tanta oscuritade?' E quelli mi dicea in parole volgari: 'Non dimandare più che utile ti sia.'

(*Vn* XII. 4–5)

Characteristically, Dante proceeds by way both of mystification and of clarification. Esoteric in tone and substance, the image none the less illuminates with superlative economy the central theme of return, of movement by degrees from the marginal and contingent to the necessary and all-embracing.[29] Dante is as yet distant. He is as yet ignorant and apprehensive with respect to what lies ahead. He is not yet home ('tu autem non sic'). Eventually, the situation of the poet as recalled in memory will coincide with that of the one who, as author, sees and understands perfectly. The wayfarer will have moved into the centre of the circle, at which point all anguish and presentiment will be resolved in joy and certainty. But for the moment all is *un*certainty, the Cavalcantian chapters of the *Vita nuova* (XIII–XVI) marking the phase of anguished irresolution preceding the new emphases of the praise style to come. Love, Dante feels, is hopelessly ambiguous. On the one hand it holds out the prospect of perfect happiness. On the other, it brings nothing but anxiety. The sight of his lady, indeed the very thought of her, is enough to vanquish his vital spirits and to reduce him to a state of psychological enervation. Conflicting intuitions strive to rob him of all peace ('lo riposo de la vita'), and, like Cavalcanti, he goes in fear for his being as a rational creature ('tremando di paura che è nel core').[30] Even here, though, in the Cavalcantian moment *par excellence* of the *Vita nuova*, we sense something of the distinctively Dantean, something of the moral philosopher's (as distinct from the natural philosopher's) search for solutions; for where Cavalcanti had been content with representation pure and simple, with the role merely of spectator, Dante is bent on finding a way out. Not for him renunciation, abandonment to his lady's mercy; instead, a struggle—if for the moment a fruitless one—for answers, for an intelligible reconciliation of the forces at work within him:

> e se io pensava di volere cercare una comune via di costoro, cioè là ove tutti s'accordassero, questa era via molto inimica verso me, cioè di chiamare e di mettermi ne le braccia de la Pietà.
>
> (XIII. 6)

There is here a typically Dantean strenuousness. Cavalcantian irresolution, the tendency of his first friend to take refuge in the merely analytical, is not for Dante. Confronted by the cruel irony of love, represented dramatically in the *gabbo* episode of Chapter XIV, where Beatrice is among those who mock him,[31] he feels the need in some way to redefine the whole question, to set it up on a fresh ideological and psychological basis; hence the 'new matter' ('matera nuova') of the chapters immediately following, the adoption of a fresh moral premise. In the meantime, however, another debt is paid, another influence honoured.

It is only in Chapters XVII and XVIII that, on the basis of what amounts to a change in the direction of spiritual energy, Dante resolves the impasse of Cavalcantianism. The argument is worth following step by step. First comes the disclaimer of XVII, the preliminary turning away on Dante's part from the all-pervasive egoism of the Cavalcantian sonnets. Having spoken at length about myself, Dante says, that, for the moment, will do. The time has come to pass on to something nobler and more delightful to contemplate:

> Poi che dissi questi tre sonetti, ne li quali parlai a questa donna però che fuoro narratori di tutto quasi lo mio stato, credendomi tacere e non dire più però che mi parea di me assai avere manifestato, avvegna che sempre poi tacesse di dire a lei, a me convenne ripigliare matera nuova e più nobile che la passata. E però che la cagione de la nuova matera è dilettevole a udire, la dicerò, quanto potrò più brievemente.[32]

Then comes the dialogue of Chapter XVIII, as delicate dramatically as it is limpid and uncompromising dialectically. 'To what end', Beatrice's companions ask, 'do you love your lady?' 'Wherein lies your happiness?' The reply is swift and unambiguous. 'My happiness', Dante says, 'lies henceforth in those words praising my lady':

> 'Madonne, lo fine del mio amore fue già lo saluto di questa donna, forse di cui voi intendete, e in quello dimorava la beatitudine, ché era

fine di tutti li miei desiderii. Ma poi che le piacque di negarlo a me, lo mio segnore Amore, la sua merzede, ha posto tutta la mia beatitudine in quello che non mi puote venire meno . . . In quelle parole che lodano la donna mia.'

(XVIII. 4 and 6)

The emphasis is as astonishing as it is decisive; for if on the one hand we witness here a redefinition of love in terms of the substance and psychology of praise—in terms, that is, of an outgoing and self-transcending act of adoration—on the other we witness the most amazing feat of cultural appropriation, the most striking example yet in Dante of the vernacular's taking over for its own ends the classical and the Christian. First, there is the Aristotelian and Ciceronian notion of friendship as disinterested concern, as affection for affection's sake:

But it is those who desire the good of their friends for their friends' sake who are most completely friends, since each loves the other for what the other is in himself.

(Aristotle, *Ethics* viii. 3; 1156^b6-11)[33]

Although friendship is certainly productive of utility, yet utility is not the primary motive of friendship . . . we are induced to form friendships, not from a mercenary contemplation of their utility, but from that pure disinterested pleasure which results from the exercise of the affection itself.

(Cicero, *De amicitia* ix. 31)

—a line taken up by Christian authors *de amicitia spirituali*:

It is a matter, not of obligation, but of affection; it is not acquired neither does it acquire. Readily undertaken, it inspires in others the same readiness. True love is happy in itself. Its only reward lies in the object itself of affection.

(Bernard of Clairvaux, *De dil. Deo* vii; *PL* 182, col. 984)

For spiritual friendship—the only true friendship, we believe—should not be sought for any worldly advantage or extrinsic cause, but willed from within on account of its own proper dignity, so that its benefit and reward are nothing other than itself.

(Aelred of Rievaulx, *De spirit. amic.* i; *PL* 195, col. 666)

Love, therefore, contracted for the purposes of some worldly gain is worthy neither of the name nor of the honour of friendship; for friendship is its own cause, its own commodity . . .

(Peter of Blois, *De amic. christ.* iv; *PL* 207, col. 875)

Next, there is the Boethian sense of love as self-sufficiency, as a movement of the spirit indifferent to circumstance. The individual, Boethius maintains in his *Consolation of Philosophy*, should never trust chance. He should instead develop and rely on his own inner resources, for this way alone lies happiness:

'Why do you mortal men seek after happiness outside yourselves, when it lies within you? You are led astray by error and ignorance. Briefly, then, I will point you towards the centre of your chief happiness. Is there anything more precious to you than yourself? Nothing, you will say. If, therefore, you enjoy yourself, then you will be in possession both of what you will never wish to lose and of what fortune can never take away from you. And so that you might see more clearly that blessedness does not lie in the random goods of fortune, consider this: if happiness is the highest good of a rational nature, and anything that can be taken away is not the highest good, since it is surpassed by what cannot be taken away, fortune by her very mutability cannot hope to lead to happiness. Again, the man who is borne along by happiness which can at any time fail, either knows or does not know its unreliability. If he does not know it, what kind of happiness can there be in the blindness of ignorance? And if he does know it, he cannot avoid being afraid of losing what he knows can be lost, whereupon perpetual fear prevents his happiness.'

(*Consol.* II, pr. iv)

—a note struck again by Dante's number one vernacular authority, Jean de Meun of the *Roman de la rose* (5301–6; and see too 3037–40 and 3953–63):

> touz les biens que dedanz toi senz
> et que si bien les connois enz,
> qui te demeurent sanz cessier
> si qu'il ne te peuent lessier
> por fere a autre autel servise:
> cist bien sunt tien a droite guise.

Whatever of worth you have within you and which you recognize as truly your own, and which remains always with you, never taking leave of you to enter another's service, that is rightly and properly yours.[34]

THE *VITA NUOVA*

And finally there is scripture: the 'better part' motif of Luke 10:42 (for the 'che non mi puote venire meno' clause of XVIII. 4);[35] the Pauline 'new creature' passages of Romans 12: 2 and 2 Cor. 5: 17 ('Do not be conformed to this world, but be transformed by the renewal of your mind that you may prove what is the will of God, what is good and acceptable and perfect'; 'therefore if anyone is in Christ, he is a new creation; the old has passed away, behold, the new has come'); and, for the *stilo de la loda*, the psalmic 'new song' ('Rejoice in the Lord . . . sing to Him a new song', Ps. 33: 1–3; 'He put a new song in my mouth, a song of praise to our God', 40: 3; 'I will sing a new song to Thee', 144: 9; 149: 1, etc.).

Amazingly, then—for never is the consistency of the solution prejudiced by the multiplicity of its sources—Dante has drawn on the Aristotelian, on the Stoic, on the Christian mystical, and on the biblical to resolve a 'courtly' question of love. Beatrice, he maintains, is a special presence, a 'cosa nova', and as such stands to be loved, not selfishly, but for herself, for the exemplary goodness she represents. Indeed, to love her in this way is to be made over again, to be cast afresh in the mould of her excellence. And this in turn generates a new style, a new spiritual utterance, the psalmody of praise and of rejoicing. True, the externals are unchanged. Beatrice is the embodiment of virtue and beauty, and Dante looks on her rapturously. But the underlying psychology is new, for the lover, far from seeking and acquiring, from pursuing happiness on the basis of his lady's concession, submits to her in a kind of self-emptying—a profitable self-emptying to be sure, but none the less a self-emptying. Like the mystics—Richard of St Victor, for example ('For what is this going forth other than a kind of self-emptying?', *Ben. maj.* IV. xvi; *PL* 196, col. 154), or Bernard of Clairvaux ('So to lose yourself . . . that you pass through a kind of self-emptying into nothingness', *De dil. Deo* x; *PL* 182, col. 990)—he is lost in the good over and beyond self, this, however, constituting the means of intimate renewal. His only purpose now (and indispensable, here as throughout, is the Pauline formula of Ephesians 3: 18–19 and, especially, 4: 13) is to grow into the 'stature of the fullness' of his lady.

Chapter XVIII, then, is the key moment in the book, the real turning-point. All that follows, and much still is to come, is in the nature of a deepening and refining of the idea of Beatrice as, on analogy with Christ, *epiphany* or special revelation. And it *is* a question of analogy—not of symbolism or of typology (for this would be to render the *Vita nuova* a work of specifically Christian inspiration), but of analogy.[36] Beatrice, the object of Dante's praise, is *like* Christ. She is like him in her provenance, in her salvific purpose, in the perfection of her being and example, and in her claim to obedience. The first stage in the working-out of this likeness comes in Chapters XX and XXI with the sonnets *Amore e 'l cor gentil sono una cosa* and, more particularly, *Ne li occhi porta la mia donna Amore*; for, miraculously Dante maintains, his lady not only actualizes a prior 'potential' or capacity for love, but instils that potential in the first place: 'ma là ove non è in potenzia, ella, mirabilemente operando, lo fa venire' (XXI. 1; and cf. 6: 'e questo è tanto a dire quanto inducere Amore in potenzia là ove non è'). Startlingly, and indeed from a theological point of view a little shockingly, she participates in the process of creation, in the bringing about *ex nihilo* of powers and possibilities in the world here below. But the Christian analogy proper—firm, but because it *is* analogy by no means reckless—begins to emerge in Chapter XXIII with Dante's delirious premonition of his lady's death, accompanied in his imagination by the crucifixion imagery of eclipse and earthquake:

> Poi mi parve vedere a poco a poco
> turbar lo sole e apparir la stella,
> e pianger elli ed ella;
> cader li augelli volando per l'are;
> e la terra tremare . . .
>
> (*Donna pietosa e di novella etate*,
> lines 49–53)

And by Chapter XXIV it is explicit. Beatrice appears with her own John the Baptist, Cavalcanti's lady Giovanna, who, as *vox clamans* or voice crying in the desert, prepares the way. The key passage runs as follows:

> . . . io vidi venire verso me una gentile donna, la quale era di famosa bieltade, e fue già molto donna di questo primo mio amico. E lo nome di questa donna era Giovanna, salvo che per la sua bieltade, secondo che altri crede, imposto l'era nome Primavera; e così era chiamata. E

appresso lei, guardando, vidi venire la mirabile Beatrice. Queste donne andaro presso di me così l'una appresso l'altra, e parve che Amore mi parlasse nel cuore, e dicesse: 'Quella prima è nominata Primavera solo per questa venuta d'oggi; ché io mossi lo imponitore del nome a chiamarla così Primavera, cioè prima verrà lo die che Beatrice si mosterrà dopo la imaginazione del suo fedele. E se anche vogli considerare lo primo nome suo, tanto è quanto dire "prima verrà", però che lo suo nome Giovanna è da quello Giovanni lo quale precedette la verace luce, dicendo: "Ego vox clamantis in deserto: parate viam Domini."'

(XXIV. 3–4)

With this the Christian-analogical structure of the book as a whole is confirmed beyond all ambiguity. Beatrice, though by no means identified with Christ and by no means a 'type' of Christ, is none the less a Christ-like presence. Hers, like Christ's, is a coming into the poet's experience to effect through grace a process of intimate renewal, of new life through enhanced understanding and rightly ordered love, Dante for his part being called upon to acknowledge and receive the good whereby he is fashioned afresh in the image of his lady.

But this essentially Christian pattern of thought is pursued still further, for Beatrice, like Christ, is destined within the economy of the whole to die—not, certainly, to effect an atonement, but to secure the life of faith, an order of understanding and activity independent of historical contingency. First, though, before we reach the point of crisis, and following in the wake of *Donne ch'avete*, comes the moment of praise represented by *Tanto gentile e tanto onesta pare* and by *Vede perfettamente* in Chapter XXVI. Here, the psychological and the analogical (Chapters XVIII–XXIV) are suspended in favour of the contemplative as Dante ponders the beauty and virtue of his lady. She is a special creature, a miraculous showing forth of divine goodness: 'non è femmina, anzi è uno de li bellissimi angeli del cielo . . . Questa è una maraviglia; che benedetto sia lo Segnore, che sì mirabilmente sae adoperare!' (XXVI. 2); and Dante for his part—and again in a manner reminiscent of Victorine mysticism—cannot but acquiesce in the ineffability of what he sees and hears: 'Io dico ch'ella si mostrava sì gentile e sì piena di tutti li piaceri, che quelli che la miravano comprendeano in loro una dolcezza onesta e soave, tanto che ridicere non lo sapeano' (XXVI. 3; cf. Richard of St Victor, *De quat. grad. viol. char.* and *Adnot. myst.* Ps. 30; *PL* 196, col. 1217 and col. 275: 'The soul thirsts for God when she desires to experience what that inward sweetness is that inebriates the mind of man when he begins to taste and to see how sweet the Lord is'; 'for by this exaltation the soul is imbued with a marvellous and immense sweetness which no sense can appreciate nor any word explain'). Typically, then, the objective coalesces with the subjective, Cavalcantian representationalism giving way to an essentially religious, and quite un-Cavalcantian, commitment to 'movement into', to immersion in the truth as now made manifest.

But if these sonnets mark the culmination of Dante's 'praise style' in the *Vita nuova*, they do not mark the resolution of his theme in the book. It is only with the death of Beatrice (anticipated in Chapter XXII by the death of her father and more ominously in XXIII by Dante's premonition of *her* death) and with the temptation episode of the 'donna gentile' that his new understanding of love as disposition is realized beyond all doubt. First, then, comes the death of Beatrice, the moment represented by the great canzone *Li occhi dolenti*. Never mind the details, Dante says at the beginning of Chapter XXVIII; they are not what matters. What matters is their inner meaning, their implication for an understanding of love won only at the expense of sorrow and self-searching; so, for example, lines 60–8 of *Li occhi dolenti*, fraught with sadness, isolation, and a sense of guilt:

> e quale è stata la mia vita, poscia
> che la mia donna andò nel secol novo,
> lingua non è che dicer lo sapesse:
> e però, donne mie, pur ch'io volesse,
> non vi saprei io dir ben quel ch'io sono,
> sì mi fa travagliar l'acerba vita;
> la quale è sì 'nvilita,
> che ogn'om par che mi dica: 'Io t'abbandono',
> veggendo la mia labbia tramortita.

(lines 60–8)

The disposition to praise is still there; but—and this is the point—delight in the good over and beyond self turns now to desolation as the embodiment of that good, Beatrice, disappears from sight. Love as 'openness to' is not yet totally secure. It is still dependent on the image, on the historical presence. The best Dante can do for the moment is to cry out for comfort, for Beatrice's compassion in his bereavement:

> Poscia piangendo, sol nel mio lamento
> chiamo Beatrice, e dico: 'Or se' tu morta?';
> e mentre ch'io la chiamo, me conforta.
>
> (lines 54–6)

And this is the mood of the poems following in the wake of *Li occhi dolenti*, of *Venite a intender li sospiri miei*, of *Quantunque volte* (nicely nuanced, however, in its sense of Beatrice as a 'spiritual bellezza grande', there to be celebrated rather than mourned), and of the first anniversary poem *Era venuta ne la mente mia*.

But next comes the moment of challenge, the testing of Dante's new understanding of love as praise on the plane of living experience—the function of the 'donna gentile' episode recorded in Chapters XXXV–XXXIX. Here there is, as we noted earlier, a problem, for Dante of the *Convivio* understands the 'donna gentile', not as a woman, but as a symbol of philosophy. There are various reasons for this. For one thing, Dante was always concerned to ensure the maximum continuity between his works, the intelligibility of one work in terms of another (so, for instance, the 'fervida e passionata' paragraph of *Convivio* I. i. 16–19, linking the *Vita nuova* to the *Convivio* in a more or less smooth progression), and for another he wished at all costs to avoid the charge of levity (so, for instance, the 'e però me non giudicare lieve e non stabile' clause of *Convivio* III. i. 12). But, again as we noted earlier, these *Vita nuova* poems make excellent sense as they stand; for in them we witness the final triumph of the new over the old, of love as disposition over love as dependence. The drama of Chapters XXXV–XXXIX (anticipated, incidentally, by the 'angel on the tablet' episode of XXXIV) repays close attention, for it is more subtle than appears at first sight. The 'donna gentile', the woman who takes pity on Dante in his sorrow, is not simply a distraction, a subsequent and unworthy infatuation. She is a means of return, a way back to former happiness. In her Dante sees the likeness of Beatrice. Her bearing, her complexion, her good will, they are all Beatrice's:

'Avvenne poi che là ovunque questa donna mi vedea, sì si facea d'una vista pietosa e d'un colore palido quasi come d'amore; onde molte fiate mi ricordava de la mia nobilissima donna, che di simile colore si mostrava tuttavia' (XXXVI. 1), and he therefore returns to her as to Beatrice. He is, in short, prey to the casuistry of the eyes, to their insidious self-recommendation as the principle of happiness; hence their chastisement, their cursing even, as instruments of forgetfulness and treachery:

Onde più volte bestemmiava la vanitade de li occhi miei, e dicea loro nel mio pensero: 'Or voi solavate fare piangere chi vedea la vostra dolorosa condizione, e ora pare che vogliate dimenticarlo per questa donna che vi mira; che non mira voi, se non in quanto le pesa de la gloriosa donna di cui piangere solete; ma quanto potete fate, ché io la vi pur rimembrerò molto spesso, maladetti occhi, ché mai, se non dopo la morte, non dovrebbero le vostre lagrime avere restate.'[37]

(XXXVII. 2)

With this, love's independence of sensible perception is at last secure, and though the soul continues to grieve, it is in the spirit of repentance and anticipation rather than of nostalgia and despair. Bereavement, no longer a stumbling block in the way of understanding, furnishes the point of departure for an ascent of the mind, for a final union of the spirit with the object of its delight, the point Dante reaches in the final, superlative, *Oltre la spera*. The argument—an argument, we must remember, which is at one and the same time confession and example—is now complete. Love as self-recovery through self-emptying has been demonstrated both intellectually and practically. It has been redefined and verified at first hand. Through it—and this, philosophically, is Dante's point of arrival—the lover transcends the hitherto impossible constraints of his own nature. He is, henceforth, 'transhumanly' (*Para.* 1. 70), in a manner open to constant self-renewal. And with this the old 'courtly' problem is itself transcended, overtaken by a new moral and existential fervour.

The final chapters of the *Vita nuova* do two things: they confirm the idea of love as a process of *becoming*, and they underline the susceptibility of Dante's theme to further development on the basis of a new and more ample philosophical culture. Important for the first of these is the pilgrimage motif of Chapter XL, serving at once to underline and, in its invitation to passers by, to universalize the central notion of journey, of being and becoming through fresh understanding (a neat reply, this, to the pilgrimage motif of the *Fiore*). Important for the second is the commitment of Chapter XLII. Dante will now devote himself, he says, to fresh study in order to celebrate Beatrice as no woman has ever been celebrated before:

E di venire a ciò io studio quanto posso, sì com'ella sae veracemente. Sì che, se piacere sarà di colui a cui tutte le cose vivono, che la mia vita duri per alquanti anni, io spero di dicer di lei quello che mai non fue detto d'alcuna. E poi piaccia a colui che è sire de la cortesia, che la mia anima se ne possa gire a vedere la gloria de la sua donna, cioè di quella benedetta Beatrice, la quale gloriosamente mira ne la faccia di colui *qui est per omnia secula benedictus.*

(XLII. 2–3)

Thus the experience of the *Vita nuova* is open-ended, open to further exploration in terms of the new scholastic learning indicated (a shade ostentatiously) by the 'e ciò dice lo Filosofo nel secondo de la Metafisica' clause of XLI. 6.

But if it is open-ended, the *Vita nuova* is at the same time a summing up, a resolution of love's ambiguities and of rhetoric's dichotomies in a new sense of the poem as spiritual-linguistic entity. And this, for all its implicit sense of their distinction, of what might be said first about love and then about style, is of the essence; for, whatever else it is, the *Vita nuova* is a meditation on the unity *in re* of form and substance, on their practical coalescence. The key chapter is XXV, worth following step by step. Dante has been taken to task, probably by one or other of the 'Tuscan' polemicists, for presenting love as a person ('sustanzia corporale') in its own right. This, he explains, is a literary trick, for love, strictly, is an 'accident within a substance', a contingency. If he speaks of it as laughing and talking, he is, like the great poets of antiquity, speaking metaphorically:

Onde, con ciò sia cosa che a li poete sia conceduta maggiore licenza di parlare che a li prosaici dittatori, e questi dicitori per rima non siano altro che poete volgari, degno e ragionevole è che a loro sia maggiore licenzia largita di parlare che a li altri parlatori volgari: onde, se alcuna figura o colore rettorico è conceduto a li poete, conceduto è a li rimatori.

(xxv. 7)

Interrogated, moreover, as to his true meaning, he is, like the great poets of antiquity, prepared to strip down his metaphors in favour of the plain sense, to reveal, as he puts it in the *Convivio* (ii. i. 3), the 'veritade ascosa sotto bella menzogna':

> E acciò che non ne pigli alcuna baldanza persona grossa, dico che né li poete parlavano così sanza ragione, né quelli che rimano deono parlare così non avendo alcuno ragionamento in loro di quello che dicono; però che grande vergogna sarebbe a colui che rimasse cose sotto vesta di figura o di colore rettorico, e poscia, domandato, non sapesse denudare le sue parole da cotale vesta, in guisa che avessero verace intendimento. E questo mio primo amico e io ne sapemo bene di quelli che così rimano stoltamente.

(xxv. 10)

And with this—the most important of Dante's glossatorial interventions in the *Vita nuova*—we are in immediate contact with the book's leading idea, with its principal area of interest; for keen as it is to unfold and to celebrate a new 'intelletto d'amore', a new understanding of love, and anxious as it is to display and to vindicate a new way of making poetry, a sweet new style, the *Vita nuova* is concerned above all to show how substance and form come together in the poem to produce a single organism, a living amalgam of signified and signifying. True, old habits of mind linger on, and Dante will continue to speak in terms of vestment and true meaning ('vesta . . . verace intendimento'). But this, for the practitioner if not for the analyst in him, is a pattern of thought now superseded; for form, with respect to content, is no mere addition, no mere embellishment of the idea in its 'plain' expression. Rather, it is the principle of its being, that whereby it actually *is* as an object of awareness. No style, no idea. No *stil novo*, no *vita nuova*. The one exists in and through the other.

2

The Mature *Rime* and the *Convivio*

(i)

The rime

WHAT happened to Dante after the death of Beatrice and the writing of the *Vita nuova* is in part recorded in and in part inferable from the *Convivio*, composed in the early years of his exile (1304–8). We know from *Convivio* II. xii, and could in any case have guessed from the *Vita nuova*, that in his bereavement he sought comfort from philosophy, from the works in particular of Boethius (the *Consolation*) and of Cicero (the *De amicitia*). At first, he says, he found them hard going, but with patience and the little grammar he had, he made his way into them, supplementing his private reading with thirty months' or so attendance at the open debates of the Dominicans and Friars Minor: 'E da questo imaginare cominciai ad andare là dov'ella [i.e. philosophy] si dimostrava veracemente, cioè ne le scuole de li religiosi e a le disputazioni de li filosofanti. Sì che in picciol tempo, forse di trenta mesi, cominciai tanto a sentire de la sua dolcezza, che lo suo amore cacciava e distruggeva ogni altro pensiero' (II. xii. 7).[1] At the same time, in the years between 1295 and 1298, he assumed a degree of 'domestic and civic responsibility' (the 'cura familiare e civile' of *Convivio* I. i. 4), thus combining the contemplative with the active, the quietude of *stilnovismo* with the practicalities of politics and of family management. The result is a fresh order of poetry designed to do two things: (*a*) to extend the substance and integrity of the 'praise style' to the celebration of philosophy, and (*b*) to develop certain key themes of that praise style, themes such as *gentilezza* and *leggiadria*, in a more rigorously dialectical fashion. Having in other words left behind him the 'fervent and passionate' years of the *Vita nuova* (*Conv.* I. i. 16), Dante embarks on the 'temperate and virile' season of the moral and pedagogical poems, on the harsher way of the 'poet of rectitude' (*poeta rectitudinis*, *DVE* II. ii. 8).

Taking first the allegorical *rime* in celebration of philosophy, there are five poems to consider: the first two canzoni of the *Convivio* (*Voi che 'ntendendo il terzo ciel movete* and *Amor, che ne la mente mi ragiona*), the ballad *Voi che savete ragionar d'Amore*, and the two sonnets *Parole mie che per lo mondo siete* and *O dolci rime che parlando andate*. Thanks to references both in the poems themselves and in the *Convivio* we can determine their order of composition. First comes *Voi che 'ntendendo*, according to *Parole mie* (lines 1–4) the first of the poems to celebrate 'quella donna in cui errai', and then the ballad *Voi che savete*, which to go by Dante's remarks in *Convivio* III. ix. 2 must have preceded *Amor, che ne la mente mi ragiona*, this, therefore, coming third in the cycle. Next comes *Parole mie*, a poem lamenting the arduousness of Dante's new love, and then *O dolci rime*, a recantation or palinode with respect to *Parole mie*. To these we must add a further group of *possibly* allegorical poems, the 'pargoletta' poems (*I' mi son pargoletta bella e nova*, *Perché ti vedi giovinetta e bella*, and *Chi guarderà già mai sanza paura*), and the two canzoni *Amor, che movi tua vertù da cielo* and *Io sento sì d'Amor la gran possanza*.

Setting aside for a moment the allegorical aspect of these poems, they all fit comfortably within the sphere of Dante's activity as a *stilnovista*. Indeed, they may be said to mark a new peak of that activity, a fresh triumph of the 'stilo de la loda'. *Voi che 'ntendendo*, it is true, reflects more closely the anguished style of Cavalcanti; on the one hand stands the 'donna gentile', as wise and formidable as she is kind and courteous, while on the other stands the thought of Beatrice in glory, the poet himself being caught up in a conflict of loyalties resolved only gradually and painfully in favour of the former. But *Amor, che ne la mente mi ragiona*, like *Donne ch'avete*, is an essay in pure praise. Solemn, measured, perfectly controlled in point of rhythm, sound, and syntax, it oscillates gently between the object of perception and the intimate sensation of knowing, enjoying, and becoming:

> Amor, che ne la mente mi ragiona
> de la mia donna disïosamente,
> move cose di lei meco sovente,
> che lo 'ntelletto sovr'esse disvia.
>   Lo suo parlar sì dolcemente sona,

> che l'anima ch'ascolta e che lo sente
> dice: 'Oh me lassa, ch'io non son possente
> di dir quel ch'odo de la donna mia!'
>
> . . . .
>
>   Cose appariscon ne lo suo aspetto
> che mostran de' piacer di Paradiso;
> dico ne li occhi e nel suo dolce riso,
> che le vi reca Amor com'a suo loco.
>   Elle soverchian lo nostro intelletto,
> come raggio di sole un frale viso;
> e perch'io non le posso mirar fiso,
> mi conven contentar di dirne poco.
>   Sua bieltà piove fiammelle di foco,
> animate d'un spirito gentile
> ch'è creatore d'ogni pensier bono;
> e rompon come trono
> li 'nnati vizii che fanno altrui vile.
> Però qual donna sente sua bieltate
> biasmar per non parer queta e umile,
> miri costei ch'è essemplo d'umiltate!
>
>   (lines 1–8 and 55–70)

Everything is there: his sense of love as inward discursiveness ('dittar dentro'), her beauty and goodness, his timidity, her grandeur, his sense of being remade in the presence of his lady, her ineffability—everything central to the praise style of Guinizzelli, and before him, to that of the Sicilians (typically Sicilianizing, for example, are the 'come raggio di sole un frale viso' of line 60 and the 'piove fiammelle di foco' of line 63). And what goes for *Amor, che ne la mente mi ragiona* goes also for the other poems of this group, for the modest praise style of *I' mi son pargoletta*, for the more sustained but at the same time more highly wrought canzoni *Amor, che movi tua vertù da cielo* and *Io sento sì d'Amor la gran possanza*, as well as for the 'durezza della donna' poems *Perché ti vedi* and *Chi guarderà già mai*, all of them a faithful restatement of familiar 'courtly' and stilnovistic themes.

But the fact of allegory sets these poems apart from their stilnovistic predecessors, marking them out as the product of a fresh phase of poetic experimentation. That they *are* allegorical—allegorical in the sense of *originally* allegorical (as distinct from praise poems subsequently *glossed* allegorically)—seems certain from their distinctive intonation. Take, for example, the following lines from *Voi che 'ntendendo* (lines 46–52) and from *Amor, che ne la mente* (lines 9–13, 23–9, and 59–62), predicable certainly of a woman but redolent rather more of the cognitive:

> 'Mira quant'ell'è pietosa e umile,
> saggia e cortese ne la sua grandezza,
> e pensa di chiamarla donna, omai!
> Ché, se tu non t'inganni, tu vedrai
> di sì alti miracoli adornezza,
> che tu dirai: "Amor, segnor verace,
> ecco l'ancella tua; fa che ti piace."'

> E certo e' mi conven lasciare in pria,
> s'io vo' trattar di quel ch'odo di lei,
> ciò che lo mio intelletto non comprende;
> e di quel che s'intende
> gran parte, perché dirlo non savrei
> . . . .
> Ogni Intelletto di là su la mira,
> e quella gente che qui s'innamora
> ne' lor pensieri la truovano ancora
> quando Amor fa sentir de la sua pace.
> Suo esser tanto a Quei che lel dà piace,
> che 'nfonde sempre in lei la sua vertute
> oltre 'l dimando di nostra natura.
> . . . .
> Elle soverchian lo nostro intelletto,
> come raggio di sole un frale viso;
> e perch'io non le posso mirar fiso,
> mi conven contentar di dirne poco.

to which we might add these lines from the first of the 'pargoletta' poems (*I' mi son pargoletta*, lines 4–10):

> I' fui del cielo, e tornerovvi ancora
> per dar de la mia luce altrui diletto;
> e chi mi vede e non se ne innamora
> d'amor non averà mai intelletto,
> ché non mi fu in piacer alcun disdetto
> quando Natura mi chiese a Colui
> che volle, donne, accompagnarmi a vui.

True, there is little here, in the words of one modern commentator, to *compel* an allegorical interpretation.[2] Everything Dante says about philosophy he has already said of

Beatrice. But the tone even so tends towards the esoteric, towards the act of understanding; so, for example, the 'see how wise and courteous she is' motif of *Voi che 'ntendendo* or the 'truth transcending my understanding' motif of *Amor, che ne la mente*, each of which—especially when taken in conjunction with the 'saranno radi | color che tua ragione intendan bene' passage of *Voi che 'ntendendo* (53–5)—suggests a moment of pure intellection. And this leads on to the importance of these poems within Dante's poetic itinerary as a whole, for apart from what they have to say about the difficulties of his experience as a poet-philosopher, as one struggling to come to grips with the complexities of metaphysics (*Conv.* IV. i. 8),[3] they witness to a new conception of allegory. Gone is the straight personification of, say, the *Fiore*, with its interminable cast of Chastity, Jealousy, Fear, Shame, Foulmouth, False Seeming, and Reason. All this gives way to a more properly analogical sense of the way in which this or that object of perception may be said to embody, and in its embodying to make known, the truth by which it is transcended. Consider, then, the following lines from *Amor, che ne la mente mi ragiona* (27–32 and 37–44):

> Suo esser tanto a Quei che lel dà piace,
> che 'nfonde sempre in lei la sua vertute
> oltre 'l dimando di nostra natura.
> La sua anima pura,
> che riceve da lui questa salute,
> lo manifesta in quel ch'ella conduce
>
> . . . . . .
>
> In lei discende la virtù divina
> sì come face in angelo che 'l vede;
> e qual donna gentil questo non crede,
> vada con lei e miri li atti sui.
> Quivi dov'ella parla, si dichina
> un spirito da ciel, che reca fede
> come l'alto valor ch'ella possiede
> è oltre quel che si conviene a nui.

where the transcendent (the 'Quei che lel dà', the 'oltre 'l dimando di nostra natura', the 'virtù divina', the 'spirito da ciel', the 'oltre quel che si conviene a nui' of the poem) is known and celebrated in and through the immanent (the ' 'nfonde sempre in lei', the 'che riceve da lui', the 'lo manifesta', the 'In lei discende', the 'ella possiede'), in and through the data of sense experience. The move is important, for, notwithstanding Dante's commitment in *Convivio* II. i. 4 to the alternativistic 'way of the poets', it means the eclipse of allegory proper in favour of something closer to historical symbolism, to a validation of the image by the truth in which it participates—a procedure decisive for the *Commedia*.

Post-dating the allegorical rime, though probably not by much, are the *rime petrose* and the Forese *tenzone*, essays respectively in the harsh and in the comic style.[4] The model and inspiration for the *petrose* lay in the poets of Provence, and in the work in particular of Arnaut Daniel (*fl.* 1180–1200), a nobleman from Périgord whose verses even in their own time were recognized as 'hard to understand' ('no son leus ad entendre').[5] But they are not hard for hard's sake; on the contrary, the difficulty of form is almost invariably the difficulty of thought and feeling, of a desperate and relentless state of mind. Typical are the following stanzas (the first and fifth) from his poem *Sols sui qui sai lo sobr'afan qe.m sortz*, predecessor to Dante's *Al poco giorno e al gran cerchio d'ombra* in the tradition of the *oda continua* or seamless stanza:

> Sols sui qui sai lo sobr'afan qe.m sortz
> Al cor, d'amor sofren per sobr'amar,
> Car mos volers es tant ferms et entiers
> C'anc no s'esduis de celliei ni s'estors
> Cui encubic al prim vezer e puois.
> C'ades ses lieis dic a lieis cochos motz;
> Puois, qan la vei, non sai—tant l'ai—que dire.

> Jois e solatz d'autra.m par fals e bortz,
> C'una de pretz ab lieis no.is pot esgar,
> Qu.l sieus solatz es dels autres sobriers.
> Hai! si no l'ai; las, tant mal m'a comors!
> Pero l'afans m'es deportz, ris e jois,
> Car en pensan sui de lieis lecs e glotz.
> Hai Dieus! si ja.n serai estiers gauzire?

I am the only one who knows the over-anguish which wells in my heart, suffering of love through over-loving, for my desire is so steadfast and entire that it never turned nor cut loose from her whom I longed for at first sight and ever after. And ever, far from her, I say to her burning words; then, when I see her, I know not what—I've so much to say.

Joy and solace from any other appear to me false and abortive, for no woman can match her in merit, and her solace is supreme above others. Ah, if I have it not; alas, she has so cruelly caught me! Yet the anguish is to me pleasure, smiles and joy, since in thought I am greedy and avid for her. Ah God! will I ever, otherwise, have joy of her?

Characteristically, a close pattern of repetition—aural, semantic, and syntactical—gives expression to the obsessive nature of the poet's passion. A studied circularity of means conveys the idea of a restless 'pensiero dominante', form thus bearing the heaviest communicative responsibility. And it is this sense of form as communication, elaborated here in the context of uncompromising 'difficulty', that engaged Dante's sympathy and admiration at a time when, in reaction, perhaps, to the idealism of the *stil novo*, he too appears to have experienced disillusionment and frustration. The exact nature of this disillusionment, emotional or intellectual, or both, is uncertain. References in the *Commedia*, as well as in a sonnet addressed to Dante by Cavalcanti (*I' vegno 'l giorno a te 'nfinite volte*),[6] suggest a period of both moral and intellectual waywardness, but they remain vague. What is certain, however, is that this moment of spiritual difficulty coincides in the *petrose* with a fresh phase of technical experimentation stimulated by the Provençaux. Quite spectacularly, one order of tension, the psychological, finds expression in another, the stylistic, in a technical obsessiveness and structural virtuosity of which the double sestina *Amor, tu vedi ben* is a *ne plus ultra*.[7] Sound, syntax, and (especially) rhyme are all exploited to the full, all marked by the same fitful energy and angularity. But most important of all is metaphor. No longer a matter merely of transference, of a carefully contrived carrying-across or *transumptio*, it bears about it an inherent necessity, a unique and incontrovertible claim to encompass the leading idea. So, for example, the first poem of the cycle, *Io son venuto al punto de la rota*, where a sense of spiritual inertia, of numbness and sterility, is known and contemplated in the wintry images of ice, cold, frost, and doom-laden skies. Here, certainly, there is no question of contrivance, of mere rhetorical agility. On the contrary, Dante is experimenting at the limits of proper predicability:

> Levasi de la rena d'Etïopia
> lo vento peregrin che l'aere turba,
> per la spera del sol ch'ora la scalda;
>   e passa il mare, onde conduce copia
> di nebbia tal che, s'altro non la sturba,
> questo emisperio chiude tutto e salda;
>   e poi si solve, e cade in bianca falda
> di fredda neve ed in noiosa pioggia,
> onde l'aere s'attrista tutto e piagne:
> e Amor, che sue ragne
> ritira in alto pel vento che poggia,
> non m'abbandona; sì è bella donna
> questa crudel che m'è data per donna.
>   .   .   .   .   .
>   Versan le vene le fummifere acque
> per li vapor che la terra ha nel ventre,
> che d'abisso li tira suso in alto;
>   onde cammino al bel giorno mi piacque
> che ora è fatto rivo, e sarà mentre
> che durerà del verno il grande assalto;
>   la terra fa un suol che par di smalto,
> e l'acqua morta si converte in vetro
> per la freddura che di fuor la serra:
> e io de la mia guerra
> non son però tornato un passo a retro,
> né vo' tornar; ché, se 'l martiro è dolce,
> la morte de' passare ogni altro dolce.

(lines 14–26; 53–65)

The same goes for the canzone *Così nel mio parlar voglio esser aspro*, doubly interesting in that the first line explicitly registers the change of tone, the embracing of a new stylistic *asperitas* over and against the *suavitas* of what has gone before—all very 'novel and unprecedented' ('novum aliquid atque intentatum artis', *DVE* II. xiii. 13), but bound even so by the now familiar principle of justification from within, by the mutual shaping of form and substance. Substance in this case involves the notion of hostility and of resistance—though ultimately, in the paroxysmal final stanza, of submission—on the part of his lady to the lover's solicitation. But it is form that affords accessibility, that conveys in the stable categories of language the 'difficulty' of Dante's experience. Syntactically, the poem is as complex as any in his *rime*. The long, run-on periods are anarchic. The sound texture is coarse (*yrsutus*) and uncompromising in its consonantalism ('petra', 'spezzi', 'scorza', 'bruca', 'rimbalza', 'squatra', 'rezzo', 'latra', 'ferza'), and the vocabulary is violent in meaning and tonality ('spezzan', 'rodermi il core a scorza a scorza', 'forza', 'denti . . . manduca', 'ancise', 'grido', 'strida', 'colpo . . . disceso giuso', 'crudele', 'squatra', 'caldo borro', 'scudiscio e ferza', 'orso . . . scherza', 'vendicherei', 'vendetta'). Again, though, it is metaphor that seals the triumph of the *petrose*. Stones, swords, shields, whips, and clubs, and the imagery generally of force, vengeance, and pitilessness, serve to articulate a state of mind incommunicable, indeed incomprehensible, were it not for the integral suggestiveness of the metaphor. Here as before, the image is all of a piece with the movement of thought and feeling—confirmation, curiously paradoxical in the circumstances, of the succession of the *petrose* to the sweet new style of the praise poems.

No less important for the prehistory of the *Commedia* (one thinks, for example, as far as the *petrose* are concerned, of the 'rime aspre e chiocce' of *Inferno* XXXII.1) is the excursion into the 'comic' style represented by the Forese *tenzone*, also belonging to the middle years of the 1290s. On the face of it, the exchange is an exercise in mutual abuse. Forese is charged with matrimonial inadequacy, gluttony, theft, and illegitimacy, while Dante is taxed with indifference to the family name and with beggary and cowardice. But it would be ingenuous to take this at face value, as an essay in malice. On the contrary, not only does the Forese encounter in *Purgatorio* XXIII give the lie to this, but the scurrilous exchange has an institutional history of its own, a history confirmed in its respectability by such names as Rustico di Filippo, Meo de' Tolomei, and Cecco Angiolieri (with whom Dante very likely corresponded). Here too, then, Dante is engaged on the literary front, in trying out and perfecting the *stilus comicus*. Take, for instance, the brilliantly conceived and executed *Chi udisse tossir la malfatata*, a marvellous piece of vituperative wit:

> Chi udisse tossir la malfatata
> moglie di Bicci vocato Forese,
> potrebbe dir ch'ell'ha forse vernata
> ove si fa 'l cristallo, in quel paese.
>
> Di mezzo agosto la truovi infreddata:
> or sappi che de' far d'ogni altro mese!
> e non le val perché dorma calzata,
> merzé del copertoio c'ha cortonese . . .
>
> La tosse, 'l freddo e l'altra mala voglia
> non l'addovien per omor ch'abbia vecchi,
> ma per difetto ch'ella sente al nido.
>
> Piange la madre, c'ha più d'una doglia,
> dicendo: 'Lassa, che per fichi secchi
> messa l'avre' 'n casa del conte Guido!'

Put alongside this Forese's hopelessly longwinded *L'altra notte mi venne una gran tosse*, with its interminable 'and then . . . and then', and we see at once, in Dante, the craftsman, the professional dealer in words, images, and sounds. Not for him the lameness of '-osso' and '-aggio', but the adventurousness of '-ecchi' and '-ido', sounds as tricky to rhyme as they are efficient in conveying the poem's essential bite. Not for him the bathos of Forese's concluding 'tornai a dietro, e compie' mi' vïaggio', but the scintillating 'messa l'avre' 'n casa del conte Guido' ('I'd have done a darn sight better pairing her off with Count Guido'), a superbly witty conclusion to a superbly witty sonnet; all in all, then, yet a further essay in stylistic possibility, in the exploring and honing to perfection of a technical register.

With this we come to the moral *rime*, to the first-fruits of Dante's career as a *poeta rectitudinis*. Dante himself explains their circumstances. They are, he says (at the beginning of *Convivio* IV), a product of Philosophy's indifference to him in the more abstruse areas of metaphysics, an indifference, and indeed at times cruelty, already registered in the allegorical *rime*. For the time being, and *faute de mieux*, he will address himself to the more pragmatic issues of moral philosophy, of *gentilezza* and of *leggiadria*. In retrospect, though, it is possible to gloss the matter more positively, for what we have here is the emergence of Dante the *moral* philosopher, of Dante the celebrant, not only of *beauty*, but of *virtue*. The terms of the problem, together with its solution, are set out in the exquisite sonnet *Due donne in cima de la mente mia*, a useful prelude to the more weighty dialectic to follow. The sonnet runs as follows:

> Due donne in cima de la mente mia
> venute sono a ragionar d'amore:
> l'una ha in sé cortesia e valore,
> prudenza e onestà in compagnia;
>
> l'altra ha bellezza e vaga leggiadria,
> adorna gentilezza le fa onore:
> e io, merzé del dolce mio signore,
> mi sto a piè de la lor signoria.
>
> Parlan Bellezza e Virtù a l'intelletto,
> e fan quistion come un cor puote stare
> intra due donne con amor perfetto.
>
> Risponde il fonte del gentil parlare
> ch'amar si può bellezza per diletto,
> e puossi amar virtù per operare.

Here, in the brief compass of fourteen lines, is the predicament of the lover turned moralist. His allegiance, Dante says, is divided between Beauty, muse of his praise poems, and Virtue, the inspiration of his new verse. Both stand commandingly at the pinnacle of his mind. But love should be undivided.[8] So what should the poet do? Submit it, he concludes, to Amore, by whom the question is gently resolved in the final tercet. There is, Amore says, no necessary contradiction between the love of Beauty and the love of Virtue, for each has its own, complementary finality ('diletto' and 'operare'), with which conscience is calmed and a new order of activity graciously inaugurated (and there is no more gracious moment in the whole of Dante's *canzoniere*).

The only two poems to have survived from this first phase of Dante's career as a moral poet (there may well have been others) are *Le dolci rime d'amor* and *Poscia ch'Amor*.[9] Their themes are nobility (*gentilezza*) and charm or grace (*leggiadria*) respectively; but Dante's point—that nobility and grace divorced from a genuine sense of the *honestum* are shallow and meretricious—is the same throughout. So, for example, the following lines (81–8 and 112–20) from *Le dolci rime*:

> Dico ch'ogni vertù principalmente
> vien da una radice:

> vertute, dico, che fa l'uom felice
> in sua operazione.
>   Questo è, secondo che l'Etica dice,
> un abito eligente,
> lo qual dimora in mezzo solamente,
> e tai parole pone.
> . . . . . .
> Però nessun si vanti
> dicendo: 'Per ischiatta io son con lei',
> ch'elli son quasi dei
> quei c'han tal grazia fuor di tutti rei;
> ché solo Iddio a l'anima la dona
> che vede in sua persona
> perfettamente star: sì ch'ad alquanti
> che seme di felicità sia costa,
> messo da Dio ne l'anima ben posta.

or these (lines 96–114) from *Poscia ch'Amor*:

> Al gran pianeto è tutta simigliante
> che, dal levante
> avante—infino a tanto che s'asconde,
> co li bei raggi infonde
> vita e vertù qua giuso
> ne la matera sì com'è disposta:
>   e questa, disdegnosa di cotante
> persone, quante
> sembiante—portan d'omo, e non responde
> il lor frutto a le fronde
> per lo mal c'hanno in uso,
> simili beni al cor gentile accosta;
>   ché 'n donar vita è tosta
> co' bei sembianti e co' begli atti novi
> ch'ognora par che trovi,
> e vertù per essemplo a chi lei piglia.
> Oh falsi cavalier, malvagi e rei,
> nemici di costei,
> ch'al prenze de le stelle s'assimiglia!

But what really matters about these poems, over and above their extension of the 'courtly' to the civic (and, as far as Dante himself is concerned, the forging of a link between youth and maturity), is their argumentative sophistication, their regard for the niceties of scholastic procedure. Typically, therefore, the main proposition is prefaced by an antithesis and complemented by a detailed development of the leading idea (the scholastic *responsio*, taking the form in *Le dolci rime* of a meditation on nobility as manifest in youth, maturity, age, and seniority). If, moreover, we add to this (*a*) the consequentialism of Dante's language ('ché . . . ché . . . però che . . . né . . . onde convien . . . sì come . . . dunque', etc.); (*b*) his philosophical lexis ('abito eligente . . . suo subietto . . . d'uno effetto . . .'); (*c*) his itemization of the argument *à la* Thomas Aquinas (whose presence as a model in Dante's mind is confirmed by the 'Contra-li-erranti mia' motif of line 141); and (*d*) the assertiveness of the poems ('falsi li riprovo . . . Dico . . . dico . . . Vedete', etc.), then we have yet another first for the vernacular lyric, a still further expansion of vernacular poetic means. True, Guittone had come some way along the same road, authorizing Dante's own departure. But Guittone, strictly, pre-dates the mingling of vernacular lyricism and scholastic logic. Dialectically, he was a primitive. Here, by contrast, the moral canzone comes of age. Conceptually secure and scientifically structured, it prepares the way both for the *Convivio* and for the *Commedia*.

With these two doctrinal poems we may take the exilic canzoni *Tre donne intorno al cor mi son venute* and *Doglia mi reca ne lo core ardire*, belonging to the years 1302–4.[10] Chronology and the experience of exile are important, for these *rime* witness to a change in the assumptions underlying Dante's activity as a 'poet of rectitude'. Gone, for the moment at least, is his confidence in the efficacy of education, in the role of the teacher in changing and fashioning men's minds. Lonely and isolated, all he can do now is to ponder the spectacle of human nature wilfully astray. The result is twofold: the searing anger of *Doglia mi reca*, an indictment as fierce as any in Dante of greed and illiberality, and the contained but painfully insistent grief of *Tre donne*.

Tre donne falls into two parts. The first is a colloquy between Amore, who dwells in the poet's heart, and the exiled virtues of justice, consisting of Drittura or the Divine and Natural Law, and of her descendants, the *jus gentium* or law of nations, and the *lex humana* or positive law. Drittura, the most dejected of the three,[11] reveals her identity and that of her companions, telling of their birth and of their sojourn with man in Eden. Since then they have lived lives of poverty and neglect, and have come seeking refuge with their kinsman, Love, who, moved by compassion, takes them in and encourages them in the hope of a new and more righteous generation to come. The tone is one of intense pathos, all the more compelling for its dignity and restraint. The vocabulary is that of suffering ('dolente', 'dolesi', 'lagrimosa', 'dolor', 'sospiri', 'trista', 'vergogna', 'pianger'), and the frequent heptasyllables, with their adjacent rhyme-sounds, point up the rhythm of sorrow. The syntax preserves throughout an unaffected purity, while binary expressions of the form 'dolente e sbigottita' ('discacciata e stanca', 'cui vertute né beltà non vale', 'in ira ed in non cale', 'discinta e scalza', 'pietoso e fello', 'a panni ed a cintura', 'doglia e vergogna') deepen and differentiate the emotional content:

> Ciascuna par dolente e sbigottita,
> come persona discacciata e stanca,
> cui tutta gente manca
> e cui vertute né beltà non vale.
> Tempo fu già nel quale,
> secondo il lor parlar, furon dilette;
> or sono a tutti in ira ed in non cale.
>
> . . . . . .
>
> Dolesi l'una con parole molto,
> e 'n su la man si posa
> come succisa rosa:
> il nudo braccio, di dolor colonna,
>   sente l'oraggio che cade dal volto;
> l'altra man tiene ascosa
> la faccia lagrimosa:
> discinta e scalza, e sol di sé par donna.
>
> (lines 9–15, 19–26)

The second part of the poem, inaugurated by the stupendous 'E io . . .' of line 73, takes up Justice's theme and explores it 'from within', from the point of view of one who has himself lived out the complex psychology of exile, its alternating courage and despair. First, the act of defiance: come what may, exile in the company of Justice is a privilege. But defiance at once gives way to nostalgia and yearning. Exile, Dante says, would weigh lightly were it not for Florence, the 'bel segno' still of his imagination. Yet Florence, he goes on, in a still further inflection of mood, has brought him to the brink of death. Now he is spent. His strength has ebbed. All that remains is the hope of forgiveness and reconciliation:

> E io, che ascolto nel parlar divino
> consolarsi e dolersi
> così alti dispersi,
> l'essilio che m'è dato, onor mi tegno:
>   ché, se giudizio o forza di destino
> vuol pur che il mondo versi
> i bianchi fiori in persi,
> cader co' buoni è pur di lode degno.
>   E se non che de gli occhi miei 'l bel segno
> per lontananza m'è tolto dal viso,
> che m'àve in foco miso,
> lieve mi conterei ciò che m'è grave.
> Ma questo foco m'àve
> già consumato sì l'ossa e la polpa
> che Morte al petto m'ha posto la chiave.
> Onde, s'io ebbi colpa,
> più lune ha volto il sol poi che fu spenta,
> se colpa muore perché l'uom si penta.
>
>   (lines 73–90)

Again, the pathos is profound, but the eloquence, and indeed the triumph, of *Tre donne* lies not so much here, in its pathos, as in the interweaving or 'telescoping' of its two main sections. The private drama of Dante in exile is enfolded by the universal drama of justice in exile, so that what the former gains in emblematic value, as a sign of the times, the latter gains in subjective intensity. The result is a statement immeasurably greater than the sum of its parts, morally analytical and yet at the same time profoundly tragic.

Tragedy too—certainly if by 'tragedy' we mean the forlorn reversal of all that should be—is at the heart of *Doglia mi reca*, except that here its mood is angry and censorious. The basic theme is liberality, *liberalitade*, though in the event the poem settles on its opposite, *cupidigia*, as its chief point of focus. This, for Dante, is a theme bristling with paradox, with the strange but inevitable truth that those who think they possess everything in truth possess nothing; for wealth is a condition of the spirit. It is a state of mind, a quality of awareness. To suppose otherwise is to reach out vainly into eternity, to dissipate energy in an endless welter of desire. The theme is a commonplace of Christian moral literature and is expounded by Dante in the fourth book of the *Convivio*: 'Quello veramente de la ricchezza è propriamente crescere, ché è sempre pur uno, sì che nulla successione quivi si vede, e per nullo termine e per nulla perfezione' (IV. xiii. 2). But here in *Doglia mi reca*, in the context of what amounts, not to sweet persuasion, but to prophetic denunciation, it takes on something of the terrifying, of the frighteningly insane:

> Corre l'avaro, ma più fugge pace:
> oh mente cieca, che non pò vedere
> lo suo folle volere
> che 'l numero, ch'ognora a passar bada,
> che 'nfinito vaneggia!
>
> (lines 69–73)

Reaching for the universe, Dante says, insisting at every turn on the absurd irony of it all, the miser lays hold of nothing; aspiring to lordship, he falls into slavery; in search of peace, he condemns himself to restlessness; desperate to be happy, he is prey to gloom and frustration:

> Chi è servo è come quello ch'è seguace
> ratto a segnore, e non sa dove vada,
> per dolorosa strada;
> come l'avaro seguitando avere,
> ch'a tutti segnoreggia.
>
> . . . .
>
> ché da sera e da mane
> hai raunato e stretto ad ambo mano
> ciò che sì tosto si rifà lontano.
>
> Come con dismisura si rauna,
> così con dismisura si distringe:
> questo è quello che pinge
> molti in servaggio
>
> . . . .
>
> ah com poca difesa
> mostra segnore a cui servo sormonta!
>
> (lines 64–8, 82–8, 97–8)

And so it goes on, Dante stating and restating, and with an at times almost incredible ferocity, the notion of wealth as tyranny, as perpetual enslavement. But there is more, for at the heart of *Doglia mi reca* is a profound sense of the times out of joint, of a painful dislocation of, in particular, beauty and virtue. Ideally, they are complementary, the one entailing and confirming the other. But virtue especially—and this is what unites *Doglia mi reca* and *Tre donne* as the poetry of exile—is adrift. Everywhere the object of scorn and indifference, she, like justice in *Tre donne*, is an outcast, condemned to a wandering life of pain and humiliation. Men, Dante says, or rather the beasts that inhabit men's likeness (line 23), are resolute. Try as she will, she cannot win them over. All her efforts are in vain:

> Fassi dinanzi da l'avaro volto
> vertù, che i suoi nimici a pace invita,
> con matera pulita,
> per allettarlo a sé; ma poco vale,
> ché sempre fugge l'esca.
>
> (lines 106–10)

Let women, therefore, bearers of beauty as men are ideally of virtue, likewise contract out. Let them too deny what is theirs to give (lines 19–21). At least then there would be a proper symmetry, a grim balance of moral and aesthetic disfigurement.

A second set of post-exilic *rime* takes up once more, and with somewhat surprising conclusions, the old stilnovistic question of love. There are two poems to consider: *Io sono stato con Amore insieme*, part of an exchange with Cino da Pistoia dating to either 1305 or 1306 and accompanied by some correspondence; and the 'mountain-song' or *montanina* canzone *Amor, da che convien pur ch'io mi doglia* (a little later, 1307). Dante's friendship with Cino went back a long way, probably to their days in Bologna where Cino was a law student. Politically, Dante must have benefited greatly from his friend's wisdom in this area, some of his own emphases (his position, for example, on the Donation of Constantine) having a distinctly Cinian ring about them. Poetically, however, and in everything to do with the theory and practice of love, it was the other way round, Cino being far and away the greater beneficiary; hence, on Cino's side, the sonnet *Dante, quando per caso s'abbandona*, simply the latest in a series of requests for advice and clarification.[12] Hope, he says, has failed him. The beauty which discourses within ('quel piacer che dentro si ragiona'—a nice variation on one of Dante's lines) has brought him to death's door. But now there is the prospect of new love and happiness. What does Dante, as one who has been both 'in and out' (a reference both to love and to exile), think? Should he proceed with his new love or not? Elsewhere, this sort of thing gets short shrift. Inclined to see Cino as fickle in matters of love, Dante taxes him with disingenuousness and with the need to bring his deeds more in line with his words:

> Degno fa voi trovare ogni tesoro
> la voce vostra sì dolce e latina,
> ma volgibile cor ven disvicina,
> ove stecco d'Amor mai non fe' foro.
>
> (*Degno fa voi trovare ogni tesoro*,
> lines 1–4)
>
> Chi s'innamora sì come voi fate,
> or qua or là, e sé lega e dissolve,
> mostra ch'Amor leggermente il saetti.
>
> Però, se leggier cor così vi volve,
> priego che con vertù il correggiate,
> sì che s'accordi i fatti a' dolci detti.
>
> (*Io mi credea del tutto esser partito*,
> lines 9–14)

Here, however, in *Io sono stato*, Cino's enquiry elicits a sympathetic response affirming love's ineluctability. I, Dante says, have laboured under love's sway for many a long year. No one knows better than I the futility of struggling against it, the hopelessness of attempting to will and to reason it away. Do, therefore, what you must:

> Io sono stato con Amore insieme
> da la circulazion del sol mia nona,
> e so com'egli affrena e come sprona,
> e come sotto lui si ride e geme.
>
> Chi ragione o virtù contra gli sprieme,
> fa come que' che 'n la tempesta sona,
> credendo far colà dove si tona
> esser le guerre de' vapori sceme.

Però nel cerchio de la sua palestra
liber arbitrio già mai non fu franco,
sì che consiglio invan vi si balestra.

Ben può con nuovi spron punger lo fianco,
e qual che sia 'l piacer ch'ora n'addestra,
seguitar si convien, se l'altro è stanco.

—a theme taken up and developed in the accompanying letter 'to a Pistoian exile' (*Ep.* III. 3): 'Since, then, the appetitive faculty, which is the seat of love, is a faculty of the sensitive soul, it is manifest that after the exhaustion of the passion by which it was brought into operation it is reserved for another.'[13] Cino must have been well pleased, but for Dante the argument is curiously out of character; for where in, say, *Donne ch'avete intelletto d'amore* or *Oltre la spera* love is associated with the intellectual soul, here instead Dante reverts to the Cavalcantian view of love as a passion of the sensitive soul and to the familiar dialectic of determinism, pain, and anxiety. And it is in this context that we have to read the last and in many ways the most searching of Dante's great love canzoni, the *montanina*.[14] Its circumstances are set out in another accompanying letter, addressed this time to Count Moroello Malaspina, with whom Dante is known to have been in contact in the latter part of 1306. In this letter he explains the cause of his recent silence, a matter, he says, not of indifference or of neglect, but of his having fallen victim once more to love, the manner of which he will, in the *montanina*, make clear. The relevant part of the letter is worth quoting in full:

It befell, then, that after my departure from the threshold of that court (which I have since so yearned for), wherein, as you so often remarked with amazement, I was privileged to be enrolled in the service of liberty, no sooner had I set foot by the streams of the Arno, in all security and heedlessness, than suddenly—oh woe is me!—like a flash of lightening from on high, a woman appeared, I know not how, in all respects answering to my inclinations both in character and in appearance. Oh how I was dumbfounded at the sight of her! But my astonishment gave place before the terror of the thunder that followed; for just as in our everyday experience the thunderclap instantly follows the flash, so, at the sight of the blaze of this beauty, Love, terrible and imperious, straightway laid hold on me. And he, raging like a despot expelled from his fatherland, who returns to his native soil after long exile, slew or expelled or fettered whatsoever within me was opposed to him. He slew, then, the praiseworthy resolve which held me aloof from women; and he pitilessly banished as suspect those unceasing meditations wherein I used to ponder the things of heaven and of earth; and finally, that my soul might never again rebel against him, he fettered my free will, so that it behoves to turn not whither I will, but whither he wills. Love, therefore, reigns within me, with no restraining influence . . .

(*Ep.* IV. 2)

Who the woman was, indeed *whether* she was in anything more than an imaginary or visionary form (some have argued in favour of a fresh vision of Beatrice), we do not know; but that is not what matters, for the *montanina* has to do, not with specific encounters, but with the *idea* of love, with what love is and how it takes hold of the suffering spirit. Indeed, what the poem does is to focus afresh, and with unprecedented power, on one in particular of the great truths of the 'courtly' and stilnovistic tradition: on the notion that the lover, far from being a *victim* of love, connives at his own subjection. Far from being a casualty of the 'amor, ch'a nullo amato amar perdona' (*Inf.* v. 103), he engineers his own suffering. So, for example, lines 19–21 and 31–40:

> L'anima folle, che al suo mal s'ingegna,
> com'ella è bella e ria,
> così dipinge, e forma la sua pena;
>
> . . . .
>
> La nimica figura, che rimane
> vittorïosa e fera
> e signoreggia la vertù che vole,
> vaga di se medesma andar mi fane
> colà dov'ella è vera,
> come simile a simil correr sòle.
> Ben conosco che va la neve al sole,
> ma più non posso: fo come colui
> che, nel podere altrui,
> va co' suoi piedi al loco ov'egli è morto.

Here, then, in the notion of the will as party to its own demise, lies, if not the novelty exactly, then the overriding concern of the *montanina*, the fresh emphasis it brings to a theme of, on the face of it, clear-cut stilnovistic (or, more exactly, Cavalcantian) inspiration. So what are we to make of the poem? A reprise of adolescent conceits, a return in circumstances of enforced idleness to the models of yester-year? By no means; the enquiry is too searching for this, too painfully self-confrontational. Rather, it is a question of Dante's taking up again the central problem of his experience as a poet and as a moralist, the problem of the relationship of love and free will, of desire and reason. The 'praise' poems, it is true, had provided an answer, but the problem, as distinct from this or that solution, survives to tax the poet's mature conscience and to provide, as never before in Dante's lyric poetry, a premiss for the *Comedy*.

(ii)

The Convivio: *introduction and course of the argument*

The *Convivio* is Dante's great 'middle period' work. Boldly conceived as a commentary, or rather as a series of commentaries, on his own poems designed to bring their substance to those 'bowed down by domestic and civic care' (I. i. 4),[15] it is an essay both in private and in social regeneration. Privately, it is a question of self-recovery in the wake of exile, of restoring the damage done to Dante's self-perception by his recent misfortune. True, Dante himself puts it in terms of self-vindication, of righting a wrong, of restoring his battered reputation in the eyes of his countrymen:

Ahi, piaciuto fosse al Dispensatore de l'universo che la cagione de la mia scusa mai non fosse stata! ché né altri contr'a me avria fallato, né io sofferto avria pena ingiustamente, pena, dico, d'essilio e di povertate. Poi che fu piacere de li cittadini de la bellissima e famosissima figlia di Roma, Fiorenza, di gittarmi fuori del suo dolce seno—nel quale nato e nutrito fui in fino al colmo de la vita mia, e nel quale, con buona pace di quella, disidero con tutto lo cuore di riposare l'animo stancato e terminare lo tempo che m'è dato–, per le parti quasi tutte a le quali questa lingua si stende, peregrino, quasi mendicando, sono andato, mostrando contra mia voglia la piaga de la fortuna, che suole ingiustamente al piagato molte volte essere imputata. Veramente io sono stato legno sanza vela e sanza governo, portato a diversi porti e foci e liti dal vento secco che vapora la dolorosa povertade; e sono apparito a li occhi a molti che forse che per alcuna fama in altra forma m'aveano imaginato, nel conspetto de' quali non solamente mia persona invilio, ma di minor pregio si fece ogni opera, sì già fatta, come quella che fosse a fare.

(*Conv.* I. iii. 3–5)

But not far beneath the surface of this, one of the most eloquent 'exile' passages in all Dante's writing, is the search for order, for the reconstruction of a persona and of a psychology shattered by the course of events. Here as before, however, the confessional aspect of the book shades off into its public aspect, its function as an instrument of collective well-being. Dante, having served his own philosophical apprenticeship, will now lay the fruits of his experience as a poet-philosopher before his chosen readers. Confident of their responsiveness to sweet persuasion—it is this which links the *Convivio* with the pre-exilic moral poems *Le dolci rime* and *Poscia ch'Amor*—he will indicate the way of properly human happiness to those immersed in civic practicality. And it is here, in its public and educational aspect, that we discover the glory and the impossibility of the *Convivio*; for, brimming as it is with concern and generosity, at the heart of the book is a characteristic ideological precariousness, a typically Dantean wavering between two views of human nature and happiness. On the one hand, there is the humanist view, the Peripatetic view, the view of man as open here and now to self-realization through understanding and moral determination. This is the positive view, the pragmatic view, the politically committed view, the view attuned to present possibility. On the other hand, there is the Christian-Platonist view, the idealist view, the mystical view even, sufficient in its sense of displacement to qualify the humanist view, to introduce an element of waiting and of melancholy. The *Convivio*, in other words, moves on two separate planes, reaching at one and the same time *out* to a particular set of socio-economic circumstances, and *up* to an esoteric order of truth and happiness. This, together with a number of other things (the wretched circumstances of its composition, its structural diffuseness, the evolutionary character of certain key themes), doubtless helped put a speedy end to the book—Dante completed little more than a quarter of what he had intended. But, as happens elsewhere in these minor works (the *De vulgari eloquentia* is the other case in point), incompleteness is of the essence; for nowhere more plainly than here do we witness the complexity of Dante's 'middle period' spirituality, the as yet unresolved struggle for philosophical peace of mind.

Strictly, Books I–III of the *Convivio* are introductory, the banquet proper getting under way in Book IV with Dante's account of nobility. Book I sets out to clear up a number of matters on the methodological front: the author's excessive self-preoccupation,[16] the difficulty of his style, and the question of language, of Dante's having opted for Italian instead of the philosophically more respectable Latin. The first two, the author's self-preoccupation and the difficulty of his style, present no problem. It is all a question, Dante explains, of his exile, of his having, where necessary with a dash of gravity ('un poco di gravezza'), to restore his tarnished reputation as a poet and philosopher:

> Onde, con ciò sia cosa che, come detto è di sopra, io mi sia quasi a tutti li Italici apprensentato, per che fatto mi sono più vile forse che 'l vero non vuole, non solamente a quelli a li quali mia fama era già corsa, ma eziandio a li altri, onde le mie cose sanza dubbio meco sono alleviate, conviemmi che con più alto stilo dea, ne la presente opera, un poco di gravezza, per la quale paia di maggiore autoritate. E questa scusa basti a la fortezza del mio comento.
>
> (I. iv. 13)

But the third, to do with the question of language, with Dante's choice of Italian instead of Latin as the means of his commentary, is trickier, Dante arguing by way of explanation from four main standpoints: order (v), obedience (vi–vii), generosity (viii–ix), and love (x–xiii). First, then, a Latin commentary to a vernacular text would have involved a disproportion or 'disconvenevole ordinazione' (v. 2), an inappropriate subordination of the greater to the lesser good; for Latin, he explains (though this, as we shall see, is a position quickly modified), is in every sense the more excellent language. It is more stable. Unlike the vernacular, which comes and goes according to usage, it is fixed from generation to generation and does not alter or decay. It is more powerful. Because of its long and distinguished tradition it has a wealth of conceptual and expressive resources unavailable to the vernacular. And finally, it is more beautiful, being in essence a work of art, carefully meditated and, unlike the vernacular, honed to perfection. There can be no question, therefore, of subordinating Latin to Italian, of using it to comment on a set of vernacular poems, for it is altogether too sovereign for this, too exalted.

The 'obedience' argument develops much the same point in terms of the master–servant image. For what, Dante asks, is the good servant? The good servant is one who serves to the letter, who does neither too little nor too much. But Latin as a servant to the vernacular would err in both ways. It would err by default in failing to serve those without Latin, and by excess in serving those without Italian—those who, despite any number of other languages they might have, have no access to the original text. Latin in this sense—and here, in the notion of psychological estrangement, of Latin's essential 'otherness' with respect to a now self-consciously Italian-vernacular habit of mind, lies the chief interest of this section—would be a stranger to the vernacular. It would be at odds with it, ignorant of its ways, and unable to serve:

> Ancora: non è conoscente de' suoi amici, però ch'è impossibile conoscere li amici, non conoscendo lo principale; onde, se non conosce lo latino lo volgare, come provato è di sopra, impossibile è a lui conoscere li suoi amici. Ancora: sanza conversazione o familiaritade impossibile è a conoscere li uomini; e lo latino non ha conversazione con tanti in alcuna lingua con quanti ha lo volgare di quella, al quale tutti sono amici; e per consequente non può conoscere li amici del volgare. E non è contradizione ciò che dire si potrebbe, che lo latino pur conversa con alquanti amici de lo volgare; ché, però non è familiare di tutti, e così non è conoscente de li amici perfettamente; però che si richiede perfetta conoscenza, e non difettiva.
>
> (I. vi. 9–11)

Next comes the argument from generosity, from 'pronta liberalitade'. All men, Dante had begun by saying in I. i. 8, are naturally friends to their neighbours, and are concerned for their well-being. But concern is a thoughtful business. It involves a sense of proportion, of what might usefully be given and usefully received; all of which in the present context points to Italian as the appropriate medium of discourse, to the vernacular in which Dante's readers ordinarily think and feel. Anything other would be surplus to requirements, unrequested, unefficacious, and ill-conceived: 'È adunque manifesto che lo volgare darà cosa utile, e lo latino non l'averebbe data' (I. ix. 9). It would be less than generous.

Finally, there is the argument from love, which, though it begins polemically, as an attack on the 'those wretched Italians who commend others' vernaculars while disparaging their own' (I. xi. I), quickly becomes a tribute to the language through which Dante has himself come about as a poet and philosopher. It was the vernacular, he says, that initiated him in the ways of art and science. It was through the vernacular that he came to Latin and to the wealth of the Latin literary and philosophical tradition (I. xiii. 5). He has, therefore, as far as Italian is concerned, much to be thankful for, and that, he says, is what in a sense his book is about. It is about seeking the welfare of a friend and acknowledging a debt of gratitude:

Anche c'è stata la benivolenza de la consuetudine; ché dal principio de la mia vita ho avuta con esso benivolenza e conversazione, e usato quello deliberando, interpetrando e questionando. Per che, se l'amistà s'accresce per la consuetudine, sì come sensibilmente appare, manifesto è che essa in me massimamente è cresciuta, che sono con esso volgare tutto mio tempo usato. E così si vede essere a questa amistà concorse tutte le cagioni generative e accrescitive de l'amistade; per che si conchiude che non solamente amore, ma perfettissimo amore sia quello ch'io a lui debbo avere e ho.

(I. xiii. 8–10)

With this, Dante moves on, in Book II, to the second stage of his apology, to an account of his own falling in love with philosophy. The text to be glossed is the pre-exilic allegorical poem *Voi che 'ntendendo il terzo ciel movete*, a miniature psychomachia or inner conflict reminiscent of Cavalcanti in its sense of drama. On the one hand, and proceeding for the moment literally, there is the thought of Beatrice in glory, a constant reminder to Dante of times past and of old allegiances:

> Suol esser vita de lo cor dolente
> un soave penser, che se ne gìa
> molte fiate a' pie' del nostro Sire,

> ove una donna glorïar vedìa
> di cui parlav'a me sì dolcemente
> che l'anima dicea: 'Io men vo' gire.'
>
> (lines 14–19)

On the other, there is the 'donna gentile', the compassionate lady of his bereavement, wise, humble, courteous, and kind (lines 46–7). The dominant mood, therefore, is one of anguish, and it is in anguish that Dante turns for sympathy and understanding to the 'movers of the third heaven':

> Voi che 'ntendendo il terzo ciel movete,
> udite il ragionar ch'è nel mio core,
> ch'io nol so dire altrui, sì mi par novo;
>   e 'l ciel che segue lo vostro valore,
> gentili creature che voi séte,
> mi tragge ne lo stato ov'io mi trovo.
>   Onde 'l parlar de la vita ch'io provo
> par che si drizzi degnamente a vui:
> però vi priego che lo mi 'ntendiate.
>
> (lines 1–9)

But who are these 'movers of the third heaven'? The answer calls for a little cosmology. The universe, Dante explains in the commentary to the poem, is made up of concentric spheres, each bearing attached to it a heavenly body: the Moon, Mercury, Venus, the Sun, Mars, Jupiter, and Saturn, with the earth in the middle. Beyond these are the firmament, comprising those stars so distant they seem to be stationary, the *primum mobile* (an object, Dante explains, of induction on the part of Ptolemy), and the Empyrean, the 'soprano edificio del mondo, nel quale tutto lo mondo s'inchiude, e di fuori dal quale nulla è' (II. iii. 11). Thus the created world is contained within the pure and limitless spirituality of the divine or 'first' mind (*Protonoè*), from which, in a process supervised by the governing Intelligences or angels, it flows forth to assume the contingent form of things here below. The key passage, worth quoting in full for its bearing on Dante's philosophical idealism in the *Convivio*, runs as follows:

E qui è da sapere che ciascuno Intelletto di sopra, secondo ch'è scritto nel libro de le Cagioni, conosce quello che è sopra sé e quello che è sotto sé. Conosce adunque Iddio sì come sua cagione, conosce quello che è sotto sé sì come suo effetto; e però che Dio è universalissima cagione di tutte le cose, conoscendo Lui, tutte le cose conosce sì, secondo lo modo de la Intelligenza. Per che tutte le Intelligenze conoscono la forma umana in quanto ella è per intenzione regolata ne la divina mente; e massimamente conoscono quella le Intelligenze motrici, però che sono spezialissime cagioni di quella e d'ogni forma generata, e conoscono quella perfettissima, tanto quanto essere puote, sì come loro regola ed essemplo. E se essa umana forma, essemplata e individuata, non è perfetta, non è manco de lo detto essemplo, ma de la materia la quale individua. Però quando dice: *Ogni Intelletto di là su la mira*, non voglio altro dire se non ch'ella è così fatta come l'essemplo intenzionale che de la umana essenzia è ne la divina mente e, per quella in tutte l'altre, massimamente in quelle menti angeliche che fabbricano col cielo queste cose di qua giuso.

(III. vi. 4–6)

The Intelligences, then, are the pure spiritual substances who, contemplating the ideas of things in the divine mind, take them up, materialize, and individualize them, remaining henceforth responsible for their proper earthly operation. And it is they, or rather the movers of the third heaven in particular—the movers presiding over the affective forces emanating from Venus—that Dante is addressing in his poem. Plaintiff in an affair of the heart, he is taking his case to the top.

But *Voi che 'ntendendo* is, as we have discovered, allegorical. It is not erotic in inspiration. It has instead to do with Dante's encounter with philosophy, with his falling in love with 'the most beautiful and virtuous daughter of the Emperor of the Universe, whom Pythagoras named Philosophy' (II. xv. 12). How, precisely? By means, Dante explains, of an analogy to be drawn between the heavens and the sciences. The Moon, for instance, is the most variable of the planets, and bears a likeness to the most changeable of the sciences—grammar. Mercury, similarly, is the most compact of the planets, and in this is like logic or dialectic. Mars, being half-way up the heavenly hierarchy, is like the proportional science *par excellence* of music, while Jupiter, the temperate planet, is like geometry, and Saturn, the most elevated of the planets, is like astrology. As for the firmament, partly seen and partly unseen, this is like physics and metaphysics, while the *primum mobile* and the Empyrean resemble respectively the architectonic or all-governing science of ethics and the perfectly still, perfectly equilibrated science of theology. Venus, then, the sweetest of the planets, is like rhetoric, the sweetest of the sciences, while her 'movers' are none other than those philosophers (Cicero and Boethius) who with honeyed words first caused Dante to love 'wisely', to take Philosophy as his lady:

> Per le ragionate similitudini si può vedere chi sono questi movitori a cu' io parlo, che sono di quello movitori, sì come Boezio e Tullio, li quali con la dolcezza di loro sermone inviarono me, come è detto di sopra, ne lo amore, cioè ne lo studio, di questa donna, gentilissima Filosofia, con li raggi de la stella loro, la quale è la scrittura di quella.
>
> (II. xv. 1)

That, then, is the meaning of *Voi che 'ntendendo*. On the face of it erotic, it is in fact esoteric, having to do with Dante's encounter, not with a woman, but with Philosophy, with the 'love of wisdom', as we are now about to learn from Book III, first and foremost in God. How, therefore, could Dante resist? Such was the persuasiveness of her eyes ('li occhi di questa donna sono le sue demonstrazioni', xv. 4) and the ravishing nature of her presence in his mind ('O dolcissimi e ineffabili sembianti, e rubatori subitani de la mente umana, che ne le mostrazioni de li occhi de la Filosofia apparite, quando essa con li suoi drudi ragiona!', ibid.) that he was caught up willy nilly in praise and adoration—the function of the great canzone *Amor, che ne la mente mi ragiona* to which we now come.

Book III, conceptually as well as expressively the most exquisite of the four, is a hymn to love; for philosophy, Dante says, rightly understood, is nothing other than the love of wisdom first and foremost in God. To love philosophy, therefore, is to love love, to be involved in the timeless and all-comprehending self-love of its author and principle. The argument runs as follows. Love, Dante explains (III. ii. 3), is a *virtus unitiva*, a tendency in the subject towards unity with the object of its perception and delight. As such it is shared by all creatures. They all have what Augustine in a chapter of the *Confessions* much cherished by Dante (XIII. ix) calls a spiritual weight or gravity (*pondus*), a natural inclination towards their proper good. Plants, for example, love vegetatively. Move them, and they will die, or else survive sadly as strangers to their environment: 'e però vedemo certe piante lungo l'acque quasi cantarsi, e certe sopra li gioghi de le montagne, e certe ne le piagge e dappiè monti; le quali, se si transmutano, o muoiono del tutto o vivono quasi triste, sì come cose disgiunte dal loro amico' (III. iii. 4). Beasts love 'sensitively', in keeping with the exigencies of the 'animal' soul, while angels, and with them man (who, uniquely, loves in all these ways), love rationally, in keeping with the exigencies of the intellectual soul. And this, to move on to the next stage of the argument, is the function, or rather the passion, of philosophy; for the term 'philosophy', though commonly used in a restrictive sense to denote this or that particular area of enquiry (physics, metaphysics, ethics, etc.), means simply 'love of wisdom', a 'friendship to understanding' ('amistanza a sapienza', III. xi. 6) characterized *efficiently* by truth (truth being what brings it about), *finally* by joy (joy being its overall purpose), and *formally* by love (love being the way in which it is known and experienced).

But now there is a problem, or if not a problem exactly, then a corollary; for if philosophy means love of wisdom, then the philosopher *par excellence* is God, for He alone is all-loving and all-knowing. In Him alone are love and knowledge as of the essence:

Filosofia è uno amoroso uso di sapienza, lo quale massimamente è in Dio, però che in Lui è somma sapienza e sommo amore e sommo atto; che non può essere altrove, se non in quanto da Esso procede. È adunque la divina Filosofia de la divina essenza, però che in Esso non può essere cosa a la sua essenzia aggiunta; ed è nobilissima, però che nobilissima è l'essenzia divina; ed è in Lui per modo perfetto e vero, quasi per etterno matrimonio. Ne l'altre intelligenze è per modo minore, quasi come druda de la quale nullo amadore prende compiuta gioia; ma nel suo aspetto contenta n'è la loro vaghezza.

(III. xii. 12–13)[17]

Man for his part is a participant, a second-order philosopher. He shares, partially and discontinuously, in the endless and endlessly fervent self-yearning (cf. *Para.* VII. 65) of his Maker. And this is Dante's point of arrival in *Convivio* III; for man, he believes, in knowing and loving as best he can, is made like God. Through contemplation he is assimilated or proportioned to his Creator:

Così dico che Dio questo amore a sua similitudine reduce, quanto esso è possibile a lui assimigliarsi. E ponsi la qualitade de la reduzione, dicendo: *Sì come face in angelo che 'l vede* . . . Dico adunque che la divina virtù sanza mezzo questo amore tragge a sua similitudine. E ciò si può fare manifesto massimamente in ciò: che, sì come lo divino amore è tutto etterno, così conviene che sia etterno lo suo obietto di necessitate, sì che etterne cose siano quelle che esso ama; e così face a questo amore amare; ché la sapienza, ne la quale questo amore fère, etterna è.

(III. xiv. 3 and 6)

Constrained through contemplation to a new order of love ('e così face a questo amore amare'), he is drawn into the experience of eternity, philosophy thus constituting the means of his homecoming, of his 'indiarsi' as Dante will say in the *Paradiso* (IV. 28), his moving into God. This, then, is the meaning of *Amor, che ne la mente mi ragiona*, a magnificent canzone, and, in this third book of the *Convivio*, magnificently interpreted. Like *Voi che 'ntendendo*, Dante maintains, it is and always was esoteric in inspiration, having to do with an order of experience inexpressible were it not for rhetorical strategy, for the equivalence *in textu* of the sweetness of his lady's voice and the sinuousness of understanding, of the nobility of her bearing and the perennial goodness of wisdom, of the limpidity of her eyes and the sweet demonstrations of reason, and of the mysterious warmth of her smile—a 'corruscazione de la dilettazione de l'anima' (III. viii. 11)—and the joy of eternal understanding.

With this Dante comes to the first course proper of his banquet, to the theory and practice of *gentilezza* or nobility. The thesis, or rather antithesis, comes from the Emperor Frederick II, who, Dante says, saw nobility as a matter of ancient wealth and pleasing manner: 'è da sapere che Federigo di Soave . . . domandato che fosse gentilezza, rispuose ch'era antica ricchezza e belli costumi' (IV. iii. 6).[18] But this, he goes on, will not do, though why not has to await an intriguing parenthesis on the nature of imperial authority. His point is that, being a legal rather than a moral authority, the emperor has no right to pronounce in matters of this kind. He is *ultra vires*. His authority runs absolutely only in matters of pure positive law, 'sì come sono le leggi de' matrimonii, de li servi, de le milizie, de li successori in dignitade' (IV. ix. 14). It is binding only in relation to the *mechanics* of society, to the kind of laws which, being a matter merely of convention or of consensus (and right, therefore, because commanded, rather than vice versa), are there to keep the wheels turning. In all other cases we are subject to him only in so far as conscience—or more precisely conscience as shaped by moral philosophy—will allow ('Altre leggi sono che sono quasi seguitatrici di natura, sì come constituire l'uomo d'etade sofficiente a ministrare, e di queste non semo in tutto subietti', IV. ix. 15). It all sounds a bit of a quibble, a bit of a technicality. But it is not, for gradually taking shape in Dante's mind (and especially eloquent in this respect is IV. vi. 18 on the ideal unity of philosophical and imperial authority 'a bene e perfettamente reggere') is the notion of the emperor and the philosopher, between them, guiding man to his specifically mortal happiness, to the kind of happiness which, entirely without clerical supervision, awaits him here and now, in this life, through the exercise of his free will. It only remains to develop the idea *intellectually*—i.e. to insist that under the dual regimen of emperor and philosopher man may attain to *speculative* happiness here and now—and to fill in the ecclesiastical side of the equation—i.e. that alongside the emperor and the philosopher, but in strict separation from them, the pope will guide man to his eternal happiness—and the solution of the *Monarchia* comes into view.

This, though, is to move on too quickly. What matters for the moment is that, even setting aside the question of jurisdiction, the emperor's remarks about *gentilezza* are not allowable. Why not? For two reasons: (*a*) because good manners, far from being a *principle* of nobility, are a *product* of it, nobility proper being something more fundamental; and (*b*) because wealth—the other main ingredient in the emperor's definition—far from bringing happiness, brings only frustration and anxiety, a point familiar to Dante from Boethius and Jean de Meun but rehearsed in the *Convivio* with tremendous verve:

Promettono le false traditrici sempre, in certo numero adunate, rendere lo raunatore pieno d'ogni appagamento; e con questa promissione conducono l'umana volontade in vizio d'avarizia. E per questo le chiama Boezio, in quello De Consolatione, pericolose, dicendo: 'Ohmè! chi fu quel primo che li pesi de l'oro coperto e le pietre che si voleano ascondere, preziosi pericoli, cavoe?' Promettono le false traditrici, se bene si guarda, di tôrre ogni sete e ogni mancanza, e apportare ogni saziamento e bastanza; e questo fanno nel principio a ciascuno uomo, questa promissione in certa quantità di loro accrescimento affermando; e poi che quivi sono adunate, in loco di saziamento e di refrigerio danno e recano sete di casso febricante intollerabile; e in loco di bastanza recano nuovo termine, cioè maggiore quantitade a desiderio, e, con questa, paura grande e sollicitudine sopra l'acquisto.

(IV. xii. 4–5)[19]

Wealth, then, or rather the getting of it, is a dead end. Or, rather, it has no end, stretching out (as we learned from *Doglia mi reca*) endlessly into infinity: 'Quello veramente de la ricchezza è propriamente crescere, ché è sempre pur uno, sì che nulla successione quivi si vede, e per nullo termine e per nulla perfezione' (IV. xiii. 2).

So what is nobility if it is not ancient wealth and pleasing manners? It is, Dante says, nothing other than man's proper perfection, the reasonableness by virtue of which he is what he is and (ideally) does what he does. The argument is set out in Chapters xvii–xxi. First, Dante says, we see around us a variety of different virtues—courage, generosity, magnanimity, and so on. But these are reducible to a common principle, to an elective habit (*habitus electivus*) tending to the middle way (IV. xvii. 7). And this in turn is reducible to the reason instilled in man by God at the moment of his conception: 'La quale, incontanente produtta, riceve da la vertù del Motore del cielo lo intelletto possibile; lo quale potenzialmente in sé adduce tutte le forme universali, secondo che sono nel suo produttore . . .' (IV. xxi. 5). True, reason differs from man to man, this, Dante says, being a question of the variables, of the 'complexion' or disposition of the stars, of the seed, and of the constituent elements. No two cases are the same. What is important, however, contingencies of this kind apart, is that nobility, by no means a matter of social or economic circumstance, is a property of human nature *as such*. It is part of what man as man actually is.

It only remains, then, to describe the *phenomena* of nobility thus understood, how it shows itself in practice, something Dante does in the rest of Book IV. In youth, he says (taking his cue by and large from Cicero),[20] nobility is a matter of grace, of vigour in mind and body, and of attentiveness to seniority. In manhood, extending from the twenty-fifth to the forty-fifth year, it is one of self-control, of strength of purpose, of courtesy, and of loyalty ('temperato e forte . . . cortese . . . leale', IV. xxvi, *passim*); while in maturity it is a question of personality's blossoming in wise and generous counsel ('e conviensi aprire l'uomo quasi com'una rosa, che più chiusa stare non puote, e l'odore, che dentro generato è, spandere', IV. xxvii. 4). Only with old age, from seventy onwards, come thoughts of immortality, of the new and qualitatively different life awaiting man hereafter. Gently, the soul lowers the sails of its ship and puts into port:

Ed è così: come lo buono marinaio, come esso appropinqua al porto, cala le sue vele, e soavemente, con debile conducimento, entra in quello; così noi dovemo calare le vele de le nostre mondane operazioni e tornare a Dio con tutto nostro intendimento e cuore, sì che a quello porto si vegna con tutta soavitade e con tutta pace.

(IV. xxviii. 3)[21]

What comes next, Dante says, in the life to come, is beyond him to describe. Indeed, it would be presumptuous, not to say superfluous, to try; for his purpose in the *Convivio*—to lay before his chosen readership the possibilities of human experience in time and space—has now been accomplished.

(iii)

The problem of the Convivio: *idealism and humanism*

The *Convivio*, said the great Dante critic Michele Barbi, is important, not for its philosophical erudition, but for what it tells us about Dante the man—about Dante the man, we might add, in the difficult years immediately after his exile. And that is correct. Replete as it is with crumbs of wisdom gleaned from beneath the philosophers' table, it testifies most eloquently, not to what Dante knew, but to what he was, to the unsettled state of a man caught between two world views. The *Commedia* presents a different picture, for though it recognizes still the main distinctions of Dante's discourse so far—reason and revelation, philosophy and theology, time and eternity, mortal and immortal happiness—there is a certain continuity between them, and thus a certain peace of mind. But for the moment all is diversity, the various elements of Dante's thought relating in terms, not of continuity, but of tension and opposition. This, then, is the line we shall now follow, looking first at the idealist aspect of Dante's argument in the *Convivio* and then at the rationalist or humanist aspect, prefacing the whole, however, with a little philosophical history.[22]

There are two ways of looking at the world: the transcendent or (in the strict philosophical sense of the term) idealist way, and the immanent or empirical way. The idealist way is represented by Plato and the Platonists. Plato was not a systematic philosopher. He was an explorer of ideas and possibilities. But, stimulated by the scepticism of the times, he was inclined to doubt the reality of the world about him. What we see, touch, and hear in the world is not the real thing. Its essence or ultimate reality lies beyond it, in a realm of pure immaterial forms or Ideas—a concept enshrined by Plato in, as it happens, the only work of his that was well known to Christian thinkers of the Middle Ages, the *Timaeus*. Everything that *is*, according to the *Timaeus*, pre-exists ideally in the First Mind or (as Dante has it) the *Protonoè* of the Divine Artificer, whence it proceeds in an orderly act of cosmic creativity, a carefully controlled coming-forth or emanation.

Now Plato's philosophical theology is productive of some very great writing, always powerfully intuitive and with an at times impressive imaginative *élan*. Metaphysically, however, it raises problems. It is not, to begin with, particularly common-sensical. It hasn't much to commend it to the kind of pragmatic soul for whom realities are realities and ideas are ideas and that's that. Also, it doesn't account very easily for abstraction and for contingency, for judgement, emotion, dimension, colour, size, weight, and so on. Do these too have an ideal form? And what of change and decay? The Idea, one imagines, is static, forever fixed in its perfection. But things down here come and go. They wax and wane. They are born and they die. How does the theory of Ideas cope with this? Only with difficulty, which is why Aristotle, who started out a Platonist, eventually took his master to task for what he came to regard as the inventiveness of it all, as the pure fiction of Ideas. By training and temperament a biologist, Aristotle was inclined, not to doubt the objects of sense experience, but to credit them with their own intrinsic reality, with their own (potentially at least) exhaustive act of being. And it was on this basis that he built his own distinctive metaphysic. Everything, he thought, has about it its own substance and accidents (so a horse, for example, is a horse and stands fifteen hands high), and its own potential and actuality (so a colt or a chrysalis has it in itself to become a stallion or a butterfly, and thus to emerge as a mature member of this or that species). But, and this is the point, these are not *real* distinctions. They are purely notional. We should not, in other words, think of horseness as independently subsistent, as living on by itself in a materially uncontaminated form. It is, on the contrary, a property of the material world, positively instantiated and qualified. This, then, is what we mean by transcendent and immanent: transcendently, it is a question with Plato of referring everything to an immaterial principle beyond—the world here below remaining merely shadowy, provisional, and incomplete—while immanently it is a question with Aristotle of crediting the world with its own actuality and dignity, with its own inherent completeness and moral worth. The over-and-beyondness of the one is met by the in-and-for-itselfness of the other.

Now this, sketchy as it is, is a necessary prelude to what happens next, in the Christian Middle Ages. For what came next was a passing to and fro, even in the midst of Christianity, between the one world view and the other. The early part of the Christian Middle Ages, through at least to the first part of the twelfth century, was predominantly Neoplatonist in outlook, the basic pattern of thought having being established by Plotinus (AD 205–70) in his formidably difficult but powerfully compelling *Enneads*. The Plotinian universe is hierarchical in character. It is like a pyramid with, at the top, the One or Platonic God, and, in immediate contiguity as the first emanation or Intelligence, the divine Mind or Logos containing the sum total of Ideas. From the Logos flows a second Intelligence or World Soul, and thereafter all the other Intelligences responsible for the modulation of the Ideas here below. Thus everything in the world around us reflects in its materiality, or rather despite its materiality, the luminosity of the Idea, the pure spiritual form from which it derives. Now this was a pattern of thought flowing down to Dante's time in two distinct traditions, non-Christian and Christian. On the non-Christian side we have (after Plotinus) Porphyry in the third century, Proclus in the fifth, and the so-called *Book of Causes* or *Liber de causis*, a version of Proclus' *Elements of Theology* translated into Latin in the twelfth century. On the Christian side—and taking care to avoid the pantheistic implications of Neoplatonism proper (everything turning out to be God)—we have Augustine in the last part of the fourth century and early part of the fifth, the Pseudo-Dionysius in the latter part of the fifth century, the (in certain moods at least) distinctly Neoplatonizing Boethius of the *Consolation* in the early part of the sixth century, and in the later Middle Ages Franciscan and even Dominican Neoplatonists such as Bonaventure and (again in certain moods) Albert the Great. Emphases vary from one to another, Platonism in Augustine, for example, coalescing with Paulinism and Pythagoreanism, in Boethius with elements of Stoicism and Peripateticism, and in Dionysius with elements of Near Eastern gnosticism, but the basic pattern remains the same; fundamental is the ultimate resolution of everything in the pure, undifferentiated spirituality of the One.

But in the middle part of the twelfth century a new mood began to settle on European philosophy. The reception of Aristotle was under way. To start with, it was a matter of his logical works, of those making up the so-called 'new logic' of the *Organon*: the *Prior* and *Posterior Analytics*, the *Topics*, and the *De elenchis* (the *Categories* and the *De interpretatione* had been translated into Latin by Boethius in the sixth century). Subsequently, however, attention turned to the ideological works, to the *Physics* and the *Metaphysics*, the *Ethics*, the *Politics*, and the *Psychology* (*De anima*), all translated and in some cases translated twice in the years between 1170 (Gerard of Cremona) and 1260 (William of Moerbeke, translator to Thomas Aquinas), and it was on this basis that the difficult task of Aristotle's rehabilitation began. At first, while it was a question merely of the *Organon*, the problems were not too acute. But as the thirteenth century wore on that changed, and Aristotle—Aristotle of, in particular, the ideological works—looked more and more of a threat; for Christian philosophy in the West had, in the way we have seen, been mainly idealist in complexion, inclined to explain the material in terms of the immaterial, the temporal in terms of the eternal, the natural in terms of the supernatural—a way of thinking tending to encourage a sceptical view of the world round about. Here, however, was Aristotle appearing for the first time in centuries to confirm the opposite, to assert, not the provisionality of the world, but its fullness, dignity, and *perseitas*—its total and incontrovertible 'in-and-for-itselfness'. What, then, was to be done about this radically subversive new antique wisdom?

There were three possibilities, with all shades in between. One possibility, adopted by the conservative Augustinians and Franciscans, was to have nothing to do with Aristotle. Aristotle, they said, was a pagan and, as far as Christians were concerned, a thoroughly bad thing. Others went to the opposite extreme. Here, they said, was a breath of fresh air, a fresh source of wisdom by no means to be contaminated or watered down by Christianity. These were the radical Aristotelians or so-called Latin Averroists, prominent among whom were Siger of Brabant (whose attitude, it appears, became progressively conciliatory), Boethius of Dacia, and latterly John of Jandun.[23] Averroes himself (1126–98), an Islamic controversialist active for most of his life in Morocco and Spain, had been over much the same ground in respect of his Neoplatonizing predecessor Avicenna (980–1037), against whom he wished to vindicate the notion of philosophy, *vis-à-vis* theology, as the pure science of the mind, non-revealed, non-imaginative, and non-popular; all of which served in turn to encourage the Latin Averroists, those living and working in the Christian West, to stress the methodological, and at key points the substantial, independence of philosophy in relation to theology. It might in other words be possible for truths excogitated philosophically, on the basis of reasonable inference, to counter truths received theologically, on the basis of revelation and of the inspired teaching of the Church, a notion for the Augustinians, or even for moderates like Aquinas, inherently impossible.

The third way was, in fact, the moderate way, represented most importantly as far as Dante was concerned by Albert the Great of Cologne (1206–80) and by his illustrious pupil and subsequent colleague Thomas Aquinas (1225–74). Albert the Great, very much to the fore among Dante's authorities in the

Convivio, was an eclectic, a collector, sifter, and purveyor of wisdom wherever it might be found. Acutely aware of what was going on, and anxious on the methodological front—in point of the dialectical processes proper to philosophy and theology respectively—to maintain the proper distinctions, he was often happy to be a Platonist and an Aristotelian at the same time. Like Dante, he was as much an enthusiast as an organizer, as much an explorer as a builder of systems. Aquinas, by contrast, was a systematist, a thinker of superlatively lucid intelligence able on the basis of a handful of leading principles to unite the conflicting claims of idealism and rationalism in a coherent whole. This in fact, the incorporation of Aristotle within the 'open-ended' perspective of Christian Neoplatonism, was his life's work, and something he accomplished (certainly to his own satisfaction and, progressively, to that of others) on the basis of *analogy*, of a sense of the way in which, given the universal hierarchy of being as proposed by the idealists, each member of that hierarchy none the less *is*, relative to its own species, absolutely. A stone, for instance, is a stone to the same degree—in the same measure of completeness—as a horse is a horse or an angel is an angel. Specifically they are different, but ontologically, in point of the extent to which (again, *secundum quid*, in respect of their own species) they *are*, they are comparable. This, clearly, is to introduce a decisive qualification into the traditional idealist way of thinking, for the *natural*, passing and ephemeral as it is, is in one sense at least—and in a fundamental sense at that—comparable with the supernatural, on equal footing with it as a type or manifestation of being (*esse*). At last, thanks to the efforts of Aquinas, the way was open for a full reception of Aristotle and for a wide-ranging review of the world in Aristotelian categories.

With this, the bare minimum for understanding what is going on in the *Convivio*, we can return to Dante; for Dante, like Albert the Great a passionate explorer of all the leading intuitions and emphases of the day, is committed at one and the same time both to the Neoplatonist way of looking at the world *and* to the Aristotelian way. Let us take first the Neoplatonist or idealist way, and proceed under this head to examine (*a*) his cosmology, (*b*) his anthropology, and (*c*) his ethic.

Dante's cosmology, as we saw from our review of the

Convivio, is 'ideal' in complexion. Whenever, in other words, he passes from the micro- to the macro-structures of the universe, he passes to the Neoplatonist pole of his inspiration, to the notion of ideal pre-existence in the First Mind of God. Bruno Nardi, referring to the *Paradiso* and to the great Neoplatonist passages in Cantos II and XIII, put it in a nutshell: 'As for the Neoplatonists so also for Dante the universe is subject to absolute spiritualization, since in the One, who is beyond all time, space, and specific determination, is the Mind, the divine Mind in which the whole of the sensible world here below, with "everything that can and cannot die", has its place.'[24] But the basic pattern of thought—the basic notion of the world here below as a perpetual out-shining of the divine essence—is there already in the *Convivio*. It is already there in, for instance, the III. vi. 4–6 passage with which we are now familiar. Let us recapitulate.[25] Dante's universe is made up of a series of concentric spheres with the earth fixed and immobile at the centre. Each sphere is material but diaphanous, and bears attached to it a heavenly body composed both of form and of matter but by nature incorruptible. (Certain of the heavenly bodies, like Venus, are attached, not to the sphere itself, but to an epicycle upon the sphere—'una speretta'—the ancient idea of epicycles enabling the astronomer to account for the apparently irregular movement of the heavens while preserving the ideal circularity of their courses.) The first seven spheres surrounding the earth are the spheres of the Moon, Mercury, Venus, the Sun, Mars, Jupiter, and Saturn, while beyond these are the 'fixed' stars making up the firmament. Now it was clear to the observer that the seven planets below the firmament enjoy two kinds of movement: (*a*) they all move about the earth on a daily basis east to west, and (*b*) they each have a contrary movement of their own west to east, which accounts for their appearing in different parts of the sky at different times of the year. The firmament too, so it was noted in the second century BC, has a contrary movement of its own west to east, at the rate of one degree per century, in the plane of the ecliptic—which means that, in keeping with what we now call the precession of the equinoxes, the sun, in its yearly journey around the earth in the ecliptic, crosses the equator a little earlier (about twenty minutes in fact) each time round. The problem about this—apart from the immensely complicated geometry of it all—was that, for reasons of a philosophical kind, the firmament (which appears now to have *two* orders of movement) could no longer be considered the first moving heaven (*primum mobile*), since philosophically everything proceeding immediately from God must, like Him, be simple. It was therefore necessary to infer an invisible first moving heaven above the firmament, and responsible *instead* of the firmament for the daily movement of the heavens around the terrestrial axis—Dante's point (after Albert the Great probably)[26] in *Convivio* II. iii. 5:

Tolomeo poi, accorgendosi che l'ottava spera si movea per più movimenti, veggendo lo cerchio suo partire da lo diritto cerchio, che volge tutto da oriente in occidente, costretto da li principii di filosofia, che di necessitade vuole uno primo mobile semplicissimo, puose un altro cielo essere fuori de lo Stellato, lo quale facesse questa revoluzione da oriente in occidente.

So far, then, so good; seven planets and the firmament or starry heaven, each with its own proper movement, but moved *en masse* by the *primum mobile* bordering on the Empyrean. The Empyrean too is invisible, constituting an article, not of reasonable inference and still less of observation, but of faith in the inspired teaching of the Church: 'Veramente, fuori di tutti questi, li cattolici pongono lo cielo Empireo . . . secondo che la Santa Chiesa vuole, che non può dire menzogna' (II. iii. 8 and 10). Curiously, the same passage seems to suggest that like the planets and the fixed stars, the Empyrean too is material ('ciò che la sua *materia* vuole'), an awkward proposition and in any case rescinded in the *Paradiso*; but however that may be (and Dante probably intended 'substance' for 'material'), its dimensions are infinite, coextensive with those of the divine mind encompassing the whole of the created universe: 'Questo è lo soprano edificio del mondo, nel quale tutto lo mondo s'inchiude, e di fuori dal quale nulla è; ed esso non è in luogo, ma formato fu solo ne la prima Mente, la quale li Greci dicono Protonoè' (II. iii. 11). This, Dante explains, is the heaven of perfect peace, the peace of divine self-knowledge (II. iii. 10). Here the elect look upon their Maker, the Empyrean thus constituting a term of desire in all intelligent creation. Indeed, this is what sets the whole thing in motion. The whirling of the *primum mobile* is the product of a fervent desire on the part of the creature to be united at every point with the Creator:

> E questo è cagione al Primo Mobile per avere velocissimo movimento; ché, per lo ferventissimo appetito ch'è 'n ciascuna parte di quello nono cielo, che è immediato a quello, d'essere congiunta con ciascuna parte di quello divinissimo ciel quieto, in quello si rivolve con tanto desiderio, che la sua velocitade è quasi incomprensibile.
>
> (II. iii. 9)

This, clearly, is where we pass from the astronomical to the cosmological, from the data of observation and necessary inference to the grand design. And here as before our key witness is the causal chain passage of III. vi. 4–6 (see p. 86 above), a conscientious restatement of Plotinian cosmology as mediated through the *Book of Causes*.[27] Strung out, Dante says, between God and the world about us are the massed ranks of separate substances, of the angels or Intelligences. Pondering in Him the intelligible forms of which He is the universal cause, they proceed to enact His creative will here below. True, not everything turns out as it should; but where it does not, this is the fault, not of the Idea or of the Intelligence, but of the matter—more or less receptive—upon which the Idea is impressed. That, though, is a detail. The basic pattern remains the same: a pyramidal pattern involving a perpetual coming-forth or raining-down of Ideas from their primary source in God, these Ideas only gradually taking on the discreteness and materiality whereby they are known and contemplated by us.

But Dante's idealist cosmology in the *Convivio* entails a no less idealist anthropology and a no less idealist ethic. It determines his conception both of man's nature and of his final end. On the Neoplatonic view, man is in essence a spirit. He is a spiritual entity caught up *pro tempore*, for the duration of his mortal life, in the flesh. He is, as Dante puts it in one of the passages we are coming to in a moment, up to his neck in matter. But all this time he longs for release, for the moment in which, free at last, he enjoys intelligible union with his Maker. Only then will he realize fully the deepest exigencies of his human nature. Now this is a pattern of thought confirmed in the *Convivio* in two ways: negatively and positively. The negative way takes the form of stressing man's imprisonment, his immersion for the time being in matter; so, for example, the 'incarceration of the soul' passage at II. iv. 17 ('mentre che l'anima è legata e incarcerata per li organi del nostro corpo'), or the following extract, to do with the opacity of human experience here and now, from II. viii. 15: 'La quale noi non potemo perfettamente vedere mentre che 'l nostro immortale col mortale è mischiato; ma vedemolo per fede perfettamente, e per ragione lo vedemo con ombra d'oscuritade, la quale incontra per mistura del mortale con l'immortale'; or this from III. vii. 5 (the 'up to the neck in matter' passage):

Così la bontà di Dio è ricevuta altrimenti da le sustanze separate, cioè da li Angeli, che sono sanza grossezza di materia, quasi diafani per la purità de la loro forma; e altrimenti da l'anima umana, che, avvegna che da una parte sia da materia libera, da un'altra è impedita, sì come l'uomo ch'è tutto ne l'acqua fuor del capo, del quale non sì può dire che tutto sia ne l'acqua né tutto fuor da quella.[28]

The idea is the same throughout. Man, Dante says, is an intellectual being. But for the moment he is an intellectual being sunk in matter, and dependent for understanding on sense perception and inference; hence his blindness and frustration, his necessary recourse to the frail edifices of observation and logic. Roll on the day when the soul, like a butterfly from a chrysalis, emerges unencumbered by the flesh into the light of perfect intellection! Only then will man be fully and unambiguously what in essence he already is—an angel-Intelligence no different from the separate substances.

This in fact—man's being in essence an angel-Intelligence—is the positive way mentioned a moment ago; for Dante, in his 'ideal' mood, departs quite radically in this matter of the soul and of what exactly it is—in his 'psychology'—from the Christian Aristotelian view represented by, say, St Thomas. Thomas too sees man as an intellectual being. He has an intellectual soul which constitutes the *form* or specific principle of his psycho-physical being. But here, in the notion of psycho-physicality, lies the difference, for while the intellectual soul of the angels is by nature independent of the flesh—a separate substance—man's soul is inherently dependent on it. It is not merely 'caught up' in the flesh. It is properly and indispensably conjoined with it, this being the condition of man's activity as man:

The intellectual soul of man occupies in the natural order of things the lowest place among intellectual substances; for, unlike the angels, it is not endowed with a natural and instinctive awareness of the truth, but has, rather, to gather this from the discrete objects of sense perception.

(*ST* Ia. 76. 5 resp.)

It is a question, therefore, as far as man and the angels are concerned, not of degree, but of kind. Man is a different sort of creature—intellectual yes, but intellectual-in-the-flesh. He is a creature naturally dependent on the body, their union being as of the essence (*ST* Ia. 75, 7 ad 3). Not so for Dante. For him, men and angels are all of a piece. They are the same kind of being, continuous and contiguous rather than separate and diverse. So, for example, III. vii. 6: 'però che . . . tra l'angelica natura, che è cosa intellettuale, e l'anima umana non sia grado alcuno, ma sia quasi l'uno a l'altro continuo per li ordini de li gradi . . . così è da porre e da credere fermamente, che sia alcuno tanto nobile e di sì alta condizione, *che quasi non sia altro che angelo*'; or III. iii. 11: 'E per la quinta e ultima natura, cioè vera umana *o, meglio dicendo, angelica*, cioè razionale, ha l'uomo amore a la veritade e a la vertude'; or, most impressively of all, III. ii. 14:

E quella anima che tutte queste potenze comprende, e perfettissima di tutte l'altre, è l'anima umana, la quale con la nobilitade de la potenza ultima, cioè ragione, participa de la divina natura a guisa di sempiterna intelligenzia; però che l'anima è tanto in quella sovrana potenza nobilitata e dinudata da materia, che la divina luce, come in angelo, raggia in quella.

Man, Dante is saying, is a threefold creature ('tripliciter spirituatus' as he puts it in the *De vulgari eloquentia*); he exists vegetatively (like the plants), animally (like the beasts), and rationally (like the angels), a powerful combination which, when perfectly realized, makes him, not a little lower, but a little higher than the angels: 'Certo da dovvero ardisco a dire che la nobilitade umana, quanto è da la parte di molti suoi frutti, quella de l'angelo soperchia, tutto che l'angelica in sua unitade sia più divina' (IV. xix. 6). The point is clear enough. Here as before, but positively, this time, rather than negatively—in terms, that is to say, of parity with the angels rather than of temporary misfortune—Dante wishes to stress man's essential spirituality, the dignity of his properly angelic nature.

And this in turn brings us to the moral and (to use the technical term) 'eudemonistic' or 'happiness' aspect of the matter. What, given the notion of man as a spiritual being captive *pro tempore* to the flesh, is his last end, his greatest good, his highest happiness? Clearly, the freeing of the soul from its corporeal prison, and its being able to make its way unhindered into the immediate presence of its Maker. The idea finds expression in various ways and in a number of different passages and contexts, but extracts such as the following are typical:

E però che naturalissimo è in Dio volere essere—però che, sì come ne lo allegato libro si legge, 'prima cosa è l'essere, e anzi a quello nulla è',—l'anima umana essere vuole naturalmente con tutto desiderio; e però che 'l suo essere dipende da Dio e per quello si conserva, naturalmente disia e vuole essere a Dio unita per lo suo essere fortificare.

(III. ii. 7)

E la ragione è questa: che lo sommo desiderio di ciascuna cosa, e prima da la natura dato, è lo ritornare a lo suo principio. E però che Dio è principio de le nostre anime e fattore di quelle simili a sé (sì come è scritto: 'Facciamo l'uomo ad imagine e similitudine nostra'), essa anima massimamente desidera di tornare a quello.

(IV. xii. 14)

The soul, Dante is saying, dependent as it is on God for its being, yearns to be reunited with Him, to root its own contingent being in the absolute and unambiguous being of its Maker. But—and with this we come to happiness proper—the joy which this reunion with God promises belongs not to this life, but to the next life. The happiness of this life, be it speculative or active, is merely 'as it were'—a 'quasi' happiness. It is substantial, but incomplete, enfolding within itself the promise of greater things to come:

E così appare che nostra beatitudine (questa felicitade di cui si parla) prima trovare potemo quasi imperfetta ne la vita attiva, cioè ne le operazioni de le morali virtudi, e poi perfetta quasi ne le operazioni de le intellettuali virtudi. Le quali due operazioni sono vie espedite e direttissime a menare a la somma beatitudine, la quale qui non si puote avere, come appare pur per quello che detto è.

(IV. xxii. 18)

The final—paradoxical—position, then, is one of sadness and disquiet. Radiant in his capacity for understanding, man is as yet a wanderer. Exalted above all creatures in the complexity and splendour of his constitution, he is far from home. All he can do is to wait, to live patiently in anticipation of the 'more-than-human' (II. viii. 6), of the qualitative otherness of eternity.

But this, as we have said, is only one side of the story, for the *Convivio* testifies, not only to Dante's idealism, but to his humanism, to the 'immanent' or Peripatetic in him as well as to the 'transcendent' or Platonist. Let us return, then, to the anthropological aspect of the question (the cosmological remaining relatively undisturbed). As we have seen, Platonist anthropology involves the notion of man as an immaterial principle of seeing, understanding, and willing immersed for the time being in the flesh. Aristotelian anthropology, by contrast, rests on the notion of man as a single psycho-physical entity. The soul's being in the body is not an unfortunate accident. It is of the essence, its union with the flesh constituting its proper *modus operandi* and dignity. Thomas Aquinas puts it well. Whereas for the Platonists, he says, the soul is like a pilot in a ship, like a mover or navigator in respect of something that, properly speaking, is not part of him, for the Aristotelians the soul is (to use the jargon) the *form* of the body, that whereby the total psycho-physical entity actually is and operates:

> if one supposes, as the Platonists did, that the intellectual soul is united to the body, not as its form, but merely as a mover, it would then be necessary to say that there is in man another substantial form through which the body, moved by the soul, is constituted in its own being. But if . . . the intellectual soul is united to the body as its substantial form, then that apart there can be no other substantial form in man.

(*ST* Ia. 76. 4 resp.)

And this, in his Peripatetic mood, is Dante's view. Man, he says in IV. xxi. 2, is made up both of soul and of body, the soul being the *act* of the body, the principle of its being and doing: 'Ove è da sapere che, sì come dice lo Filosofo nel secondo de l'Anima, l'anima è atto del corpo' (III. vi. 11). There is an intrinsic relationship between the soul and the body. It is a question, not of captivity and unwilling servitude, but of ontological and operational necessity; so, for example, the following lines from the beginning of *Convivio* III. iii (2–5), with their twin emphasis on the complexity of human nature, at once mineral, vegetative, animal, and rational, and on the substantial unity of that nature, on its formal irreducibility:

Onde è da sapere che ciascuna cosa, come detto è di sopra, per la ragione di sopra mostrata ha 'l suo speziale amore. Come le corpora simplici hanno amore naturato in sé a lo luogo proprio—e però la terra sempre discende al centro, lo fuoco ha amore naturato a la circunferenza di sopra, lungo lo cielo de la Luna, e però sempre sale a quello—, le corpora composte prima, sì come sono le minere, hanno amore a lo luogo dove la loro generazione è ordinata, e in quello crescono e a quello devono vigore e potenza; onde vedemo la calamita sempre da la parte de la sua generazione ricevere vertù. Le piante, che sono prima animate, hanno amore a certo luogo più manifestamente, secondo che la complessione richiede; e però vedemo certe piante lungo l'acque quasi cantarsi, e certe sopra li gioghi de le montagne, e certe ne le piagge e dappiè monti; le quali, se si transmutano, o muoiono del tutto o vivono quasi triste, sì come cose disgiunte dal loro amico. Li animali bruti hanno più manifesto amore non solamente a li luoghi, ma l'uno l'altro vedemo amare. Li uomini hanno loro proprio amore a le perfette e oneste cose. E però che l'uomo—*avvegna che una sola sustanza sia tutta sua forma per la sua nobilitade*—ha in sé natura divina, queste cose e tutti questi amori puote avere, e tutti li ha.

In fact, the passage has as much of the Augustinian about it as the Aristotelian. It is yet another reminiscence of Dante's favourite chapter XIII. ix (the 'everything has its own spiritual weight' chapter) of the *Confessions*. But what interests us for the moment is its Aristotelian aspect, its sense of man's unity in diversity. Radically differentiated in his make up, he is none the less one in substance. He is not an amalgam but a single essence variously endowed with the faculties of 'living, sensing, and reasoning' (III. ii. 11), all part and parcel of one and the same organic entity.

Now this too is a view of human nature carrying implications on the moral and eudemonistic front; for given its insistence on the psycho-physicality of human nature, on the union of mind and body in a single organism (a view, we have to appreciate, tending to locate human activity proper *this side* of death—which is why Aristotle is not much concerned with what happens next), it calls for an end commensurate, not with what man *will be*, but with what he is here and now—a being ontologically in the flesh. And that end, Dante now sets about demonstrating, is an *active* end, the kind of end associated, not so much with *seeing* and *discovering*—that is the speculative end—as with *doing* and *determining*. The key passage here is IV. ix. 4–7, and it runs as follows:

E a vedere li termini de le nostre operazioni, è da sapere che solo quelle sono nostre operazioni che subiacciono a la ragione e a la volontade, che se in noi è l'operazione digestiva, questa non è umana, ma naturale. Ed è da sapere che la nostra ragione a quattro maniere d'operazioni, diversamente da considerare, è ordinata: che operazioni sono che ella solamente considera, e non fa né può fare alcuna di quelle, sì come sono le cose naturali e le sopranaturali e le matematice; e operazioni che essa considera e fa nel proprio atto suo le quali si chiamano razionali, sì come sono arti di parlare; e operazioni sono che ella considera e fa in materia di fuori di sé, sì come sono arti meccanice. E queste tutte operazioni, avvegna che 'l considerare loro subiaccia a la nostra volontade, elle per loro a nostra volontade non subiacciono; ché, perché noi volessimo che le cose gravi salissero per natura suso, e perché noi volessimo che 'l silogismo con falsi principii conchiudesse veritade dimostrando, e perché noi volessimo che la casa sedesse così forte pendente come diritta, non sarebbe; però che di queste operazioni non fattori propriamente, ma li trovatori semo. Altri l'ordinò, e fece maggior fattore. Sono anche operazioni che la nostra ragione considera ne l'atto de la volontade, sì come offendere e giovare, sì come star fermo e fuggire a la battaglia, sì come stare casto e lussuriare, e queste del tutto soggiacciono a la nostra volontade; e però semo detti da loro buoni e rei perch'elle sono proprie nostre del tutto, perché, quanto la nostra volontade ottenere puote, tanto le nostre operazioni si stendono.

Man, Dante says, is involved in various kinds of activity. Some of them, like physics and mathematics, entail no more than looking and considering. Here, we are mere 'finders' ('trovatori') or spectators. Others require something to be done, a step to be taken, a decision to be made, and it is by virtue of what we *do*, as distinct from what we *discover*, that we are deemed good or bad. Doing, then, rather than finding is where we are most at home. Acting rather than contemplating is where we are at our most authentically human, and, accordingly, at our happiest:

E queste [Dante is speaking of the eleven species of moral virtue listed in the passage immediately preceding] sono quelle che fanno l'uomo beato, o vero felice, ne la loro operazione, sì come dice lo Filosofo nel primo de l'Etica quando diffinisce la Felicitade, dicendo che 'Felicitade è operazione secondo virtude in vita perfetta'.

(IV. xvii. 8)[29]

But that is not all, for—and here we come to the crux of the matter, to the (from a specifically Christian point of view) *punctum dolens* of Dante's argument hereabouts in the *Convivio*— not only has man a distinctly 'human' (as opposed to 'more-than-human' or angelic) order of activity here and now, but he also has a perfectly 'human' authority: Aristotle. Aristotle, Dante says, gifted as he was with a virtually divine intellect ('ingegno singulare e quasi divino'), has said it all. Look no further for a complete, infallible, and universal guide to mortal happiness:

Altri furono, e cominciamento ebbero da Socrate e poi dal suo successore Platone, che, agguardando più sottilmente, e veggendo che ne le nostre operazioni si potea peccare e peccavasi nel troppo e nel poco, dissero che la nostra operazione sanza soperchio e sanza difetto, misurata col mezzo per nostra elezione preso, ch'è virtù, era quel fine di che al presente si ragiona; e chiamaronlo 'operazione con virtù'. E questi furono Academici chiamati, sì come fue Platone e Speusippo suo nepote; chiamati per luogo così dove Plato studiava, cioè Academia; né da Socrate presero vocabulo, però che ne la sua filosofia nulla fu affermato. Veramente Aristotile, che Stagirite ebbe sopranome, e Zenocrate Calcedonio, suo compagnone, per lo 'ngegno singulare e quasi divino che la natura in Aristotile messo avea, questo fine conoscendo per lo modo socratico quasi e academico, limaro e a perfezione la filosofia morale redussero; e massimamente Aristotile. E però che Aristotile cominciò a disputare andando in qua e in lae, chiamati furono—lui, dico, e li suoi compagni—Peripatetici, che tanto vale quanto 'deambulatori'. E però che la perfezione di questa moralitade per Aristotile terminata fue, lo nome de li Academici si spense, e tutti quelli, che a questa setta si presero, Peripatetici sono chiamati; e tiene questa gente oggi lo reggimento del mondo in dottrina per tutte parti, e puotesi appellare quasi cattolica oppinione. Per che vedere si può Aristotile essere additatore e conduttore de la gente a questo segno.

(IV. vi. 13–16)

Theologically—indeed not only theologically but morally too—the argument is astonishing, for without so much as a nod in the direction of Christ or of things specifically Christian, Dante defers in the matter of human conduct here and now to the pagan philosopher of antiquity. And it is in the absoluteness of that referral—the 'limaro e a perfezione' and 'cattolica oppinione' clauses of the passage just quoted—that the problem lies; for while few have any misgivings about admitting the authority of Aristotle in the sphere of moral philosophy—Albert, Thomas, and even the Augustinians and Franciscans do that—to do so *absolutely* is to separate out what within the Christian conscience proper preserves an ideal unity, a reforming immanence of the new (the Christian) in the old (the pagan). It is to inhibit completely the revitalizing function of the former in respect of the latter. Revitalization gives way to alternativism, to a curious and, potentially at least, deeply destabilizing moral and psychological dualism.

Astonishing or not, however, the position is as all-pervasive in the *Convivio* as the philosophical idealism we have just looked at. Let us pause over a couple of instances. Take the ordering of the sciences episode in Book II. Dante, it will be remembered, is busy interpreting his poem *Voi che 'ntendendo il terzo ciel movete* allegorically, each heaven representing a particular science. Thus the moon represents grammar, Mercury logic, Venus rhetoric, and so on up to the firmament as representative of physics and metaphysics, the *primum mobile* as representative of ethics, and the Empyrean as representative of theology. But what is odd about this arrangement (though one might be forgiven for not noticing it first time round) is the ascendancy of ethics over metaphysics; for ethics, *ex parte obiecti* (in respect of what it is about), is merely human, merely mundane and local. Metaphysics, by contrast, is the quasi-divine science of being as being, of being as the precondition of everything else. So how can it be inferior? The answer is all of a piece with the moral humanism we have just been considering; for ethics, though *ex parte obiecti* lower than metaphysics, is *ex parte subiecti*, or from the point of view of the one who knows, higher than metaphysics; it is higher than metaphysics for the simple reason that, as the specifically human science of doing and determining, it is absolute. Lived out, so to speak, from within, it is present in its totality to the philosopher, which is more than can be said of metaphysics. In addition, therefore, to its function as the architectonic or organizational science (it is ethics, according to II. xiv. 14 ff., that directs the various kinds of human activity to their proper end), it is also delimitative. It marks out, in its completeness, the completeness of the *human* over and against the *more-than-human*, of the moral-philosophical over and against the theological.[30]

Another case of delimitation, of the marking out of the human over and against the more-than-human, is to be found in the closing 'arc of life' chapters in Book IV. Book IV, it will be remembered, is all about nobility, redefined by Dante in IV. xx. 9 as the 'seed of happiness planted by God in the well disposed soul'. Having set out all the pros and cons, he comes at last to illustrate his theme practically, to show how nobility understood in this way manifests itself in the four ages of man. The argument runs as follows. Youth ('adolescenzia'), extending up to a man's twenty-fifth year, is obedient and receptive to the wisdom of its elders; healthy in body and mind, it is modest, restrained, and socially discreet. Manhood ('gioventute'), extending from the twenty-fifth to the forty-fifth year, is characterized by grace, resolution, loyalty, integrity, and courtesy. Like the *Convivio* itself ('temperato e virile', I. i. 16), it is firm and disciplined ('temperato e forte', IV. xxvi, *passim*). Age too ('senettute') is gracious, but graciously expansive. Personality opens like a rose allowing others to enjoy its sweet scent, its rich accumulation of wisdom and experience (IV. xxvii. 4). But—and this is the point—it is only with old age ('senio'), the fourth stage of man's earthly pilgrimage, that he looks on to the next life and begins to ponder his immortal destiny. As with the mariner, Dante says (IV. xxviii. 3), who, reaching the end of his journey, lowers the sails of his ship and slips gently into port, so also we, at the end of our mortal journey, turn 'with firm and whole-hearted intent' to God. The image is exquisite, perfectly fashioned in point of imaginative finesse and Stoic equilibrium. But it is not particularly Christian. It in no way answers to the Christian sense—fully grasped by Dante of the earlier *Vita nuova*—of the inherence of the eternal in the temporal, of the new in the old. Instead, inherence is replaced by continuity, by a 'following on' of the Christian on the Stoic. Man here and now is fine. He has all he needs. Christianity, as far as this life is concerned, is supplementary, over and beyond, next in line.

Towards the end of the *Purgatorio* Dante, newly arrived in the earthly paradise and once again in the presence of Beatrice, has his lady reproach him with a number of serious misdemeanours. In the first place, he has betrayed her memory and pursued empty pleasures (*Purg.* XXX. 124–32). In the second, he has lacked resolve, abandoned his former purposes, and been a lover of sights and sounds (XXXI. 37–63). And in the third, he has toyed with false doctrines (XXXIII. 79–90). Dante, for his part, is reduced to tears, and confesses to the truth of the accusations. But what, more precisely, was he guilty of? The first two passages suggest at the very least something erotic. No sooner had Beatrice died than Dante was entangled with someone new, with a compassionate 'donna gentile' and a passing 'pargoletta'. The third, with its mention of a 'school', suggests something more intellectual—a flirting, maybe, with the kind of radical Aristotelianism represented by Boethius of Dacia or by Siger of Brabant and from time to time infecting the *Convivio*. But more probable than this—for Dantean Averroism turns out, as we shall see, to be a very mild affair—is a general sense on Dante's part of his own exaggerated philosophism, of a tendency in his middle years to exult too freely in the possibilities of nature ungraced. And this, basically, is the point; for while on the face of it the *Commedia* too has a vigorous and optimistic sense of the possibilities of nature ungraced—of what it is possible for man to accomplish through free will alone (so Dante, led by Virgil, is able through the reasonable discipline of self-knowledge and self-cleansing to recover a moral innocence)—there is a difference, nature in the *Commedia* never in truth being ungraced. Free and self-directing certainly, it presupposes even so the divine initiative, God's prior making-possible and setting-in-motion. So, for example, the plea of the three heavenly ladies, of Maria, of Lucia, and of Beatrice, to Virgil that he might embark on his saving mission; the intervention before the walls of Dis of the divine messenger; the Purgatorial ship of souls; Virgil's prayer to the Sun and the enabling light of day; the intervention of the angelic swordsmen in the valley of the princes; Dante's lifting by Lucia to Purgatory proper—they all testify to a sense of prevenience, to a sense of grace, not simply as 'elevating' and 'saving', but as preceding, as facilitating, as orientating, indeed as permeating through and through. And this bears on the *Convivio*, with its characteristic concern for the humanly self-sufficient, for the morally independent, for the human as against the more-than-human, all of which stand now to be qualified in the light of a more specifically Christian sense of moral and theological immanentism, of the inherence of grace in nature—of, in a word, the incarnational; hence the strictures of the *Purgatorio*, its sense of the *Convivio* and of what it stands for as a passing but potentially dangerous moment of half-awareness, of ill-restrained enthusiasm.[31]

We cannot leave the *Convivio*, however, without a word on three other matters—already in part touched on—linking the book to what comes next in the *Monarchia* and in the *Commedia*: the question (*a*) of Dante's Averroism, (*b*) of his political thought, and (*c*) of his language in the *Convivio*. The question of Dante's Averroism, or radical Aristotelianism, is a complicated one. As we saw earlier, the Aristotelian camp in the West was divided into two main groups (with as many positions as there were thinkers in between): the moderates, including Albert the Great and Thomas Aquinas, and the purists or integralists. The moderates wished for synthesis, for a harmonious coming to terms on the part of Christianity with the new naturalistic emphases of Aristotelianism; thus the 'natural' experience of knowing and willing here and now stood to be lifted—not to be abolished but to be lifted—to a new plane of perfection hereafter. The radicals, however, sensed in this an element of betrayal, a muddying of the pure waters of Peripateticism. Christianity, they thought, was one thing and Aristotelianism was something else, and on certain key issues (like the creation of the world or the immortality of the soul) the two might not see eye to eye. Now Dante was an admirer of Averroes. In him he saw something of his own opposition to the excesses, moral and political alike, of theologism, to the more or less indiscriminate taking up of the here and now in the hereafter. Like Averroes, he wished to vindicate the viability of human nature in time and space, the triumph of man's search for wisdom and happiness this side of death. So was Dante an Averroist or was he not? And if so, how close did he come to heresy?

The question is tricky not least in that—setting aside for the moment the extent to which Dante had actually read the Averroists—Averroes himself stood for a variety of different things. In the first place he stood for the formidably difficult doctrine of monopsychism, for the idea that man, far from having an intellect of his own, shared for the duration of his mortal life in the one universal intellect—a doctrine tending to the denial of personal immortality. In the second place, he stood for a sharp separating out of philosophy and theology, theology, he thought, representing a pre-philosophical and thus primitive form of human understanding. And in the third place, he stood for the possibility of perfect speculative happiness here and now, for the idea that some men at least might aspire to an act of total understanding this side of death.

Now as far as Dante is concerned the first two of these, monopsychism and philosophy's superiority to theology, present no difficulty. Restricting ourselves to the *Convivio* (for the whole thing comes up again in the *Monarchia*) and taking first the question of monopsychism, there is Dante's long and detailed discourse in Book IV about the genesis of the human soul. The soul, he explains (Chapter xxi), is, in respect of its vegetative and animal parts, generated *ex materia*. It derives from matter as acted on by the generative influence of the parents and by the stars, and as conditioned by the 'complexion' or constitution of the elements. But the intellectual part of the soul, the part of it bearing the imprint of the divine, is instilled into matter immediately by God. It is God's new creation:

E però dico che quando l'umano seme cade nel suo recettaculo, cioè ne la matrice, esso porta seco la vertù de l'anima generativa e la vertù del cielo e la vertù de li elementi legati, cioè la complessione; e matura e dispone la materia a la vertù formativa, la quale si è de l'anima generante; e la vertù formativa prepara li organi a la vertù celestiale, che produce, de la potenza del seme, l'anima in vita. La quale, incontanente produtta, riceve da la vertù del Motore del cielo lo intelletto possibile; lo quale potenzialmente in sé adduce tutte le forme universali, secondo che sono nel suo produttore, e tanto meno quanto più dilungato da la prima Intelligenza è.

(IV. xxi. 4–5)

The emphasis is decisive, for whatever its primary purpose here in *Convivio* IV—to show that nobility is the 'seed of happiness implanted by God in the well-disposed soul'—it excludes once and for all the possibility of Dantean monopsychism. For Dante, the intellectual soul is the form of the body in respect of each individual—a notion reaffirmed in the lovely lines 67–75 of *Purgatorio* XXV:

> Apri a la verità che viene il petto;
> e sappi che, sì tosto come al feto
> l'articular del cerebro è perfetto,
>   lo motor primo a lui si volge lieto
> sovra tant'arte di natura, e spira
> spirito novo, di vertù repleto,
>   che ciò che trova attivo quivi, tira
> in sua sustanzia, e fassi un'alma sola,
> che vive e sente e sé in sé rigira.

What, then, of philosophy in relation to theology? How far did Dante of the *Convivio* subscribe to the Averroistic notion of theology as popular myth? In a word, he didn't; for not only does philosophy stand to be corroborated by theology (so, for example, II. viii. 14 or IV. xxi. 11 ff.), but theology—according to Dante's interpretation of Solomon in the *Song of Songs* (6: 8 in the Vulgate)—emerges as the 'dove' of the sciences, as the most perfect and most perfectly peaceful of them all: 'Tutte scienze chiama regine e drude e ancille; e questa chiama colomba perché è sanza macula di lite, e questa chiama perfetta perché perfettamente ne fa il vero vedere nel quale si cheta l'anima nostra' (II. xiv. 20).

What, then, of human happiness? How far did Dante assent to the Averroistic idea that some men at least might achieve perfect happiness here and now? Here, it has to be said, there is a problem, for on this point Dante both did and did not go along with Averroes. He *did* go along with Averroes to the extent that he believed that man might accomplish perfect happiness here and now in the *active* life, but that is not what Averroes meant. Averroes meant *speculative* happiness, the possibility of perfect *intellectual* happiness in this life.[32] And this is where things become complicated, for Dante, in a curious passage in *Convivio* III. xv, speaks of the way in which, not merely *actively* but *speculatively* too, man might after all aspire to perfect happiness here and now. The passage runs as follows:

> A ciò si può chiaramente rispondere che lo desiderio naturale in ciascuna cosa è misurato secondo la possibilitade de la cosa desiderante; altrimenti andrebbe in contrario di se medesimo, che impossibile è; e la Natura l'avrebbe fatto indarno, che è anche impossibile. In contrario andrebbe: ché, desiderando la sua perfezione, desiderrebbe la sua imperfezione, imperò che desiderrebbe sé sempre desiderare e non compiere mai suo desiderio (e in questo errore cade l'avaro maladetto, e non s'accorge che desidera sé sempre desiderare, andando dietro al numero impossibile a giugnere). Avrebbelo anco la Natura fatto indarno, però che non sarebbe ad alcuno fine ordinato. E però l'umano desiderio è misurato in questa vita a quella scienza che qui avere si può, e quello punto non passa se non per errore, lo quale è di fuori di naturale intenzione . . . Onde, con ciò sia cosa che conoscere di Dio (e di certe altre cose) quello esso è, non sia possibile a la nostra natura, quello da noi naturalmente non è desiderato di sapere. E per questo è la dubitazione soluta.
>
> (III. xv. 8–10)

—a line taken up, with reference now to Averroes himself (the Commentator), in IV. xiii. 7–8:

> Ancora qui si risponde, che non è vero ciò che si oppone, cioè che mai non si viene a l'ultimo, che li nostri desiderii naturali, sì come di sopra nel terzo trattato è mostrato, sono a certo termine discendenti; e quello de la scienza è naturale, sì che certo termine quello compie, avvegna che pochi, per male camminare, compiano la giornata. E chi intende lo Commentatore nel terzo de l'Anima, questo intende da lui.

Man, Dante says, is a creature open to understanding (the 'all men naturally desire to know' clause of I. i. 1). But if understanding is not after all possible (God, the ultimate object of understanding, being a mystery), then something is clearly amiss. Nature—perish the thought—has slipped up.[33] We must therefore conclude, Dante says, that there is in man a natural term of desire, a point at which he knows all he wishes to know and is thereafter content. If by chance he discovers anything else, then that is an added bonus. The important thing is that he has reached the term of his desire and is satisfied.

What, therefore, are we to make of this? In short, very little, and for two reasons. First, the Averroistic moment of the *Convivio* is very much a 'moment'. Both in III. xv and in IV. xiii it has very much the appearance, and certainly the function, of getting Dante off the hook, of disposing quickly of an awkward objection—namely, that if philosophy cannot after all make us happy, then what use is philosophy? Secondly, it runs contrary to everything else he says on the subject. Take, for instance, this from III. vi. 7:

> Dove è da sapere che ciascuna cosa massimamente desidera la sua perfezione, e in quella si queta ogni suo desiderio, e per quella ogni cosa è desiderata; e questo è quello desiderio che sempre ne fa parere ogni dilettazione manca: ché nulla dilettazione è sì grande in questa vita che a l'anima nostra possa tôrre la sete, che sempre lo desiderio, che detto è, non rimagna nel pensiero.

or, more especially, the following lines from IV. xxii:

> E così appare che nostra beatitudine (questa felicitade di cui si parla) prima trovare potemo quasi imperfetta ne la vita attiva, cioè ne le operazioni de le morali virtudi, e poi perfetta quasi ne le operazioni de le intellettuali virtudi. Le quali due operazioni sono vie espedite e direttissime a menare a la somma beatitudine, la quale qui non si puote avere, come appare pur per quello che detto è.
>
> (IV. xxii. 18)

His drift is always the same: a degree of happiness here and now, yes, but perfect happiness, no. As creatures *pro tempore* in the flesh, we have to be patient. Our reward will come, not here, but hereafter, in the next life. Now this is by no means to diminish the strength of Dante's humanism in the *Convivio*, his commitment to the 'human' as distinct from the 'more-than-human'. But it is to stress that Dante's Averroism, as Averroism, is a very pale shadow of the original. If to be an Averroist was to be a monopsychist, then that is the end of the matter. There is nothing in Dante to suggest that he even toyed with the idea. If, however, to be an Averroist was to experiment with the possibility of perfect happiness here and now, then fine, Dante was an Averroist. But true Averroism involved a lot more than that, and there is little in Dante's exploration of man's present happiness that could not have come—with a dash, admittedly, of enthusiasm—from the *Ethics* of Aristotle.[34]

More important, however, than the Averroist issue—as it turns out a bit of a non-issue—is the political aspect of the *Convivio*, and how this ushers in the solutions of the *Monarchia*. This again is something we have already touched on, for occupying a good part of Book IV is Dante's account of the difference, but mutual necessity, of imperial authority and of philosophical authority. Philosophical authority, Dante maintains in the key sixth chapter, is there to guide man to his mortal end ('Per che vedere si può Aristotile essere additatore e conduttore de la gente a questo segno', vi. 16), while imperial authority is there to keep him in order:

> Per che, tutto ricogliendo, è manifesto lo principale intento, cioè che l'autoritade del filosofo sommo, di cui s'intende, sia piena di tutto vigore. E non repugna a la imperiale autoritade; ma quella sanza questa è pericolosa, e questa sanza quella è quasi debile, non per sé, ma per la disordinanza de la gente; sì che l'una con l'altra congiunta utilissime e pienissime sono d'ogni vigore. E però si scrive in quello di Sapienza: 'Amate lo lume de la sapienza, voi tutti che siete dinanzi a' populi.' Ciò è a dire: congiungasi la filosofica autoritade con la imperiale, a bene e perfettamente reggere.
>
> (IV. vi. 17–18)

Here, then, lies the basis for Dante's solution in the *Monarchia*, this too involving a union of imperial and philosophical authority to guide man to his mortal end and happiness.

But this is not the only way in which the *Convivio* looks forward politically; for *Convivio* IV shows every sign of the new prophetically-minded, as distinct from pedagogically-minded, Dante of the *Monarchia* and of the *Commedia*. The point is that the *Convivio* is, as we have seen, a work of philosophical enlightenment. It is a work which seeks through instruction and persuasion to initiate men in the ways of mortal happiness. But *Convivio* IV witnesses to the emergence of two new and potentially, indeed necessarily, transforming forces: Virgil and Augustine. Let us take first Virgil. The *Aeneid*, whatever else it was, was an act of celebration. Composed in the time of

Augustus, when 'mankind rested happily in universal peace' (*Mon.* I. xvi. 2), it confirmed the providential nature of Rome's ascendancy and her claim to universal dominion. Consider, for example, these lines from Book I (278–82):

> his ego nec metas rerum nec tempora pono:
> imperium sine fine dedi. quin aspera Iuno,
> quae mare nunc terrasque metu caelumque fatigat,
> consilia in melius referet, mecumque fovebit
> Romanos, rerum dominos gentemque togatam.

To Romans I set no boundary in space or time. I have granted them dominion, and it has no end. Yes, and even the furious Juno, who now wearies sea, earth, and heaven with the strain of fear, shall amend her plans, and she and I will foster the nation which wears the toga, the Roman nation, masters of the world.

or these from Book IV (229–31):

> sed fore qui gravidam imperiis belloque frementem
> Italiam regeret, genus alto a sanguine Teucri
> proderet, ac totum sub leges mitteret orbem.

No, he was to guide an Italy which is to be a breeding ground of leadership and clamorous with noise of war, transmit a lineage from proud Teucer's blood, and subject the whole earth to the rule of law.

And that, only more so, was how it read to successive generations, to those anxious in the as yet far-off centuries of the late Middle Ages to see a Roman-imperial restoration. Here in the *Aeneid* was indisputable evidence of God's high purpose for Rome. Much along the lines of the Jewish exodus, He had conducted the Trojan exiles across the sea, protected them in endless adversity, and established them in the promised land of Latium as His chosen people and future masters of the world. That at any rate was Dante's view:

Certo e manifesto esser dee, rimembrando la vita di costoro e de li altri divini cittadini, non sanza alcuna luce de la divina bontade, aggiunta sopra la loro buona natura, essere tante mirabili operazioni state; e manifesto esser dee questi eccellentissimi essere stati strumenti con li quali procedette la divina provedenza ne lo romano imperio, dove più volte parve le braccia di Dio essere presenti. E non puose Iddio le mani proprie a la battaglia dove li Albani con li Romani, dal principio, per lo capo del regno combattero, quando uno solo Romano ne le mani ebbe la franchigia di Roma? Non puose Iddio le mani proprie, quando li Franceschi, tutta Roma presa, prendeano di furto Campidoglio di notte, e solamente la voce d'una oca fé ciò sentire? E non puose Iddio le mani quando, per la guerra d'Annibale avendo perduti tanti cittadini che tre moggia d'anella in Africa erano portati, li Romani volsero abbandonare la terra, se quel benedetto Scipione giovane non avesse impresa l'andata in Africa per la sua franchezza? E non puose Iddio le mani quando uno nuovo cittadino di picciola condizione, cioè Tullio, contra tanto cittadino quanto era Catellina la romana libertà difese? Certo sì. Per che più chiedere non si dee, a vedere che spezial nascimento e spezial processo, da Dio pensato e ordinato, fosse quello de la santa cittade. Certo di ferma sono oppinione che le pietre, che ne le mura sue stanno, siano degne di reverenzia, e lo suolo, dov'ella siede, sia degno oltre quello che per li uomini è predicato e approvato.

(IV. v. 17–20)[35]

This in itself, however, need not have impinged too much on the nature and purposes of the *Convivio*, were it not for the other force beginning now to emerge in Dante's mind, the Augustinian. Now Augustine's view of Rome is ambiguous. Sensitive to the Pauline notion that all authority is from God, he (like Dante after him) sees at work in Roman history a certain providential necessity, a certain responsiveness to God's pre-established plan. But at the same time it was a violent history, a history of self-seeking and of aggression.[36] And this—stepping back for a moment to locate Rome in the context of man's universal history—served to confirm him in his characteristic sense of mankind's moral perversion generally. The key event was Eden. What happened in Eden, and indeed continues to happen wherever Christ is excluded, was catastrophic. By no means merely disadvantaged, deprived for the time being of certain original privileges (as the Christian Aristotelians tended to think), man was spoilt through and through. Henceforth, and short of God's fresh intervention in Christ, he was totally lost, hopelessly prey to his own anarchic self.[37] Now Dante is by nature a rationalist and an optimist. Man, he believes, is a creature of moral dignity, and is capable as such of enacting the goodness proper to his nature. But nagging even so at his rationalism and optimism is an Augustinian sense of man as radically astray, as lost to the wayward forces of his own being and bent on self-destruction. Take, for instance, the 'con ciò sia cosa che l'animo umano in terminata possessione di terra non si queti' clause of IV. iv. 3, or the 'se questa—cioè equitade—li uomini la conoscessero, e conosciuta servassero, la ragione scritta non sarebbe mestiere' clause of ix. 8 (a quote, explicitly, from Augustine);[38] they each reflect on what Dante is coming now to regard as the truth of the situation—the deep-rootedness of man's moral ill and his insusceptibility to reason alone. And this, to link now his Augustinianism with his Virgilianism, means a new approach to the kinds of problem raised in the *Convivio*. It means a prophetic rather than a pedagogical approach, an approach based, not on reasonable affirmation or sweet persuasion, on the honeyed words of the philosophers, but on disclosure, on the dramatic showing-forth of God's plan. It means a vatic declaration of the ends He proposes and the means—including, signally, the restored Roman Empire—He provides, a total revelation of the grand design in time and eternity. With this, we are on the threshold of the *Commedia*. The *Convivio*, the magisterial *Convivio*, is slipping into the past. Typically, it is superseded in the very moment of its composition.[39]

Finally, there is the linguistic aspect of the *Convivio*, its courageous fashioning of a vernacular language equal to the exigencies of scholastic reasoning. Take, for example, the following passage from IV. iv:

E a queste ragioni si possono reducere parole del Filosofo, ch'egli ne la Politica dice: che quando più cose ad uno fine sono ordinate, una di quelle conviene essere regolante, o vero reggente, e tutte l'altre rette e regolate. Sì come vedemo in una nave, che diversi offici e diversi fini di quella a uno solo fine sono ordinati, cioè a prendere loro desiderato porto per salutevole via, dove, sì come ciascuno officiale ordina la propria operazione nel proprio fine, così è uno che tutti questi fini considera, e ordina quelli ne l'ultimo di tutti, e questo è lo nocchiero, a la cui voce tutti obedire deono. Questo vedemo ne li religioni, ne li esserciti, in tutte quelle cose che sono, come detto è, a fine ordinate. Per che manifestamente vedere si può che a perfezione de la universale religione de la umana spezie conviene essere uno, quasi nocchiero, che, considerando le diverse condizioni del mondo a li diversi e necessari offici ordinare, abbia del tutto universale e inrepugnabile officio di comandare. E questo officio per eccellenza Imperio è chiamato, sanza nulla addizione, però che esso è di tutti li altri comandamenti comandamento. E così chi a questo officio è posto è chiamato Imperadore, però che di tutti li comandamenti elli è comandatore, e quello che esso dice a tutti è legge, e per tutti dee essere obedito, e ogni altro comandamento da quello di costui prendere vigore e autoritade. E così si manifesta la imperiale maiestade e autoritade essere altissima ne l'umana compagnia.

(IV. iv. 5–7)

The ideas are familiar. Dante is concerned with the hierarchy of authority culminating in the emperor, in the one who has control of the last end. The pattern of thought is perfectly Aristotelian, perfectly faithful to what he had discovered in the *Ethics* (I. i; 1094^{a}1 ff.). But—and this is the point—so is his style, his appropriation in the vernacular of the various linguistic devices of the Latin in which Aristotle was currently contemplated. So, for instance, the 'Sì come vedemo . . .' of line 4 (the Latin 'et sic videmus . . .'); or the 'Per che manifestamente vedere si può . . .' and 'E così si manifesta . . .' of lines 11 and 21 (the Latin 'Et sicut patet quod . . .'); or the plethora of *però che*s and *cioè*s (the Latin *sicut* and *idest*)—devices which, coupled with the equally numerous *Dico che*s, *Rispondo che*s, and *Con ciò sia cosa che*s of the text, confirm its sense of the vernacular as dialectically mature, as adept in the handling of complex ideas. True, the *Convivio* does not have about it the argumentative precision of the *Commedia*, its superb sense of rhythm and reason at one within the economy of the tercet; but here especially the triumph of the major work presupposes the initiative of the minor work, its confirmation of the vernacular as equal in expressive power to Latin ('però che si vedrà la sua vertù, sì com'è per esso altissimi e novissimi concetti convenevolemente, sufficientemente e acconciamente, quasi come per esso latino, manifestare', I. x. 12).[40]

Linguistically, therefore, as well as politically, the *Convivio* looks forward, and constitutes a point of departure for the *Commedia*, a vital part of its prehistory. But as well as anticipatory, the *Convivio* is confirmatory. As well as a point of departure, it affirms the wealth and generosity of Dante's humanity *there and then*, in the years immediately following his exile. On the wealth side, there is no need for further comment; the text, as we have seen, teems with intuitions and emphases of the most diverse kind—Platonist, Aristotelian, Stoic, and Christian—with Dante committed passionately to them all. Dialectically, this makes for confusion, for a continuous and ultimately impossible veering between one way of looking at the world and another; but it is the confusion, and indeed the impossibility of the book, that makes for its value as *testimony*, as witness to Dante's endless desire to seek out, to understand, and to celebrate. And it is at this point, at the point of discovery and celebration, that wealth overflows into generosity, into a wish to benefit the less fortunate.[41] The key passage here is the final paragraph of Book I, worth citing in full for its splendidly biblicizing crescendo:

> Così rivolgendo li occhi a dietro, e raccogliendo le ragioni prenotate, puotesi vedere questo pane, col quale si deono mangiare le infrascritte canzoni, essere sufficientemente purgato da le macule e da l'essere di biado; per che tempo è d'intendere a ministrare le vivande. Questo sarà quello pane orzato del quale si satolleranno migliaia, e a me ne soperchieranno le sporte piene. Questo sarà luce nuova, sole nuovo, lo quale surgerà là dove l'usato tramonterà, e darà luce a coloro che sono in tenebre e in oscuritade per lo usato sole che a loro non luce.
>
> (I. xiii. 11–12)[42]

Having known for himself the power of philosophy to reform, Dante proposes to share the experience, and thus to draw his readers into a firmer sense of their private and collective identity. His book was to be a gift, a spiritual banquet, a means of new life. In the event, the gift was indifferently received, and quickly enough he passed from the pedagogical to the prophetic, to the awesome moment of anagogical disclosure. But for a moment—and this, ultimately, is what matters about the *Convivio*—the idea obtained, and obtained in a spirit of total generosity.

3
The *De vulgari eloquentia*

COINCIDING chronologically with the *Convivio* is the *De vulgari eloquentia*. Dante's reference to it in *Convivio* I. v. 10, in the context of remarks to do with the evolutionary character of language, suggests a work already envisaged in its main lines ('uno libello ch'io intendo di fare') but not as yet begun. An internal reference (I. xii. 5) suggests that the first book was written not later than 1305, and the second probably followed straight away. Ostensibly the book is, or quickly becomes, an *ars poetica*, belonging to the tradition of classical and medieval *poetriae* or manuals of rhetoric and literary composition on which Dante from time to time draws in the course of his own work. To this extent it reflects again his characteristic tendency towards periodic self-interrogation. Having experimented over the years in a variety of different poetic styles—Guittonian, Cavalcantian, and Guinizzellian—and with a variety of different themes—erotic, moral, and philosophical—but, as he himself puts it, 'casually rather than systematically' ('casu magis quam arte', II. iv. 1), he turns now to the ordering of his experience and to the elaboration of its basic principles.

But the *De vulgari* does not start out as an *ars poetica*. The problem it sets out to solve, as distinct from the solution it offers, is not a literary one. It is linguistic, and rooted in the *Convivio*. This, as we have seen, is a work of philosophical philanthropy. It sets out to enlighten those 'bowed down by domestic and civic care' and to bring them on in the way of properly human happiness. It demands, therefore, as its medium of exchange, not Latin, the traditional means of philosophical discourse, but the vernacular, the language in which men and women function as creatures of deciding and doing. This, then, is Dante's first job in the *Convivio*: the vindication of Italian as equal to the business in hand.[1] But that is easier said than done, for the claims of Latin, traditionally the language of 'high culture', to superiority seem incontrovertible.

Take, for example, stability. Which, of Latin and the vernacular, is the more stable, the more long-lasting, the more immune to fashion, the more universally viable? Latin; for Latin does not come and go. It does not wax or wane. Unlike the vernacular, it is always the same, always dependable, always communicative:

> Per nobilità, perché lo latino è perpetuo e non corruttibile, e lo volgare è non stabile e corruttibile. Onde vedemo ne le scritture antiche de le comedie e tragedie latine, che non si possono transmutare, quello medesimo che oggi avemo; che non avviene del volgare, lo quale a piacimento artificiato si transmuta.
>
> (*Conv.* I. v. 7–8)

Not only this, however, but Latin is conceptually and expressively more powerful, more adequate—as one might expect after all this time—to the demands of thought and feeling. What after all, Dante asks, is language other than an articulation of thought? That is its job. And Latin, he says, as those with both languages know well, is in this respect more efficient: 'onde, con ciò sia cosa che lo latino molte cose manifesta concepute ne la mente che lo volgare far non può, sì come sanno quelli che hanno l'uno e l'altro sermone, più è la vertù sua che quella del volgare' (*Conv.* I. v. 12). In this sense too, then, Latin is superior.

Finally, there is the question of beauty, understood by Dante to mean an orderly coming together of parts within the whole ('Quella cosa dice l'uomo essere bella cui le parti debitamente si rispondono, per che de la loro armonia resulta piacimento', I. v. 13). Here too, he says, Italian is a poor second; for the vernacular is a matter of chance, of random usage. No one deliberates about it. It just happens. Latin, by contrast, is more managed, more 'artistic', and thus more pleasing: 'Dunque quello sermone è più bello ne lo quale più debitamente si rispondono le parole; e più debitamente si rispondono in latino che in volgare, però che lo volgare seguita uso, e lo latino arte' (I. v. 14). Latin, then, wins hands down. It is in every respect superior to Italian. It is nobler in that it has the virtue of stability, abler in that it is more expressive, and more beautiful in that it is technically more refined.

But as Book I of the *Convivio* goes on, the basis of Dante's argument changes. There is a shift in the criteria he uses to judge linguistic excellence. The high-cultural criterion whereby Latin inevitably emerges as the superior tongue gives way to a sense of language, or rather of the vernacular, as the natural means of communication, as that whereby individuals and communities ordinarily give shape to the substance of thought and feeling. True, Dante himself does not argue in this way. As often as not he argues quantitatively, in terms of the number of people having access to Latin or to the Italian vernacular. But emerging in the course of *Convivio* I. v–ix is a sense of the vernacular as intrinsically suitable, as inherently appropriate to a particular set of historical circumstances. And with this change in Dante's way of looking at language comes a change in his evaluation of the vernacular in relation to Latin; for Italian, though still lagging behind in point of stability, is on a par with Latin in point of expressive power and of beauty. The key passage is the following from I. x:

Ché per questo comento la gran bontade del volgare di sì si vedrà; però che si vedrà la sua vertù, sì com'è per esso altissimi e novissimi concetti convenevolemente, sufficientemente e acconciamente, quasi come per esso latino, manifestare; la quale non si potea bene manifestare ne le cose rimate, per le accidentali adornezze che quivi sono connesse, cioè la rima e lo numero regolato; sì come non si può bene manifestare la bellezza d'una donna, quando li adornamenti de l'azzimare e de le vestimenta la fanno più ammirare che essa medesima. Onde, chi vuole ben giudicare d'una donna, guardi quella quando solo sua naturale bellezza si sta con lei, da tutto accidentale adornamento discompagnata; sì come sarà questo comento, nel quale si vedrà l'agevolezza de le sue sillabe, le proprietadi de le sue costruzioni e le soavi orazioni che di lui si fanno; le quali chi bene agguarderà, vedrà essere piene di dolcissima e d'amabilissima bellezza.

(I. x. 12–13)

A fresh conception of the vernacular as the *form* or principle of private and collective consciousness leads to a fresh sense of its dignity, and this is where the *De vulgari eloquentia* begins. Prompted by the new linguistic intuitions of *Convivio* I, Dante starts out by making good the deficiency of *Convivio* I. x, by restoring the vernacular in respect, not only of expressive power and beauty, but of nobility. His point is that man, as far as this life is concerned, is a creature in the flesh, which means that, unlike the angels (who communicate by means of intuition) and unlike the beasts (who communicate by means of instinct), he needs a psycho-physical sign to impart the otherwise inscrutable content of his mind, emotion, and imagination. He needs a 'rationale signum et sensuale':

> It was necessary, therefore, for the purposes of communicating ideas one to another, for the human race to have a rational and sensible sign—rational because it has both to receive from and to transmit to the reason, and sensible because nothing passes from one reason to another save by a sensible medium. Thus if it were solely rational it could not be transferred, and if it were solely sensible it could neither receive from nor deposit in the reason.
>
> (I. iii. 2)[2]

In fact, Dante goes a step beyond this; for it is clear from the argument generally, as it will be again from Book II of the *De vulgari*, that the word, far from being merely utilitarian, is an index, and indeed a coefficient, of self-hood, that whereby the individual comes virtually to constitute a species of his own:

> Since, therefore, man is moved, not by natural instinct, but by reason, and since, in respect of discernment, of judgement, and of choice, this reason varies from individual to individual such that each appears to enjoy the status of a separate species, so to speak, in his own right, there is, we believe, no question (as in the beasts) of one man understanding another simply on the basis of what he does or of what he experiences.
>
> (I. iii. 1)

Moreover, it is through language, Dante says in I. v. 2, that we give expression to the deepest exigencies of our human nature, to the exigencies of praise and of celebration:

> But if anyone maintains the opposite view, that he had no need to speak, since he was as yet the only human being in existence and since God discerns all our secrets with no need of words, even before we do; then with all the reverence that becomes us in making judgements about His eternal will we say that, even though God did know, and indeed foreknew (which is all the same in God) the first speaker's thought without his having to speak, yet He wished him even so to speak, that in the working out of so great a gift He might be glorified who had freely given it; hence, as a matter of belief, the divine origin of

the joy we experience in exercising in an orderly way the effects of our human nature.[3]

Language, therefore, whatever its social function, has also a psychological function, a part to play in the coming about of personality, in the confirmation in this or that individual of his proper humanity. But that, for the moment, is incidental, the main thrust of Dante's argument being utilitarian, based on man's need for communication by signs. And from this utilitarian, as distinct from high-cultural, point of view, the vernacular cannot but be the nobler of the two languages; for it is through the vernacular that man customarily functions. It is through the vernacular that he perceives, articulates, and communicates the substance of what he is, the thoughts, feelings, intuitions, and judgements which both unite him with, and distinguish him from, his fellow human beings. As for Latin, it is, in this context, a mere contrivance, a mere artefact, as subordinate to the vernacular as art is to nature—the leading proposition of I. i:

But since what matters with a science is not so much proving as defining its subject, that the basis on which it rests might be more clearly understood, we shall hurry on to define the vernacular as that which children learn from those around them when they first begin to distinguish one word from another; or, more briefly, that which we acquire without any rule, by imitating our nurses. From this we have another, secondary language which the Romans called 'grammar'. This secondary language is also possessed by the Greeks and by other peoples, but not by all; few, indeed, succeed in acquiring it, since understanding and expertise in this are a matter of much time and patient study.

Of these two languages the nobler is the vernacular, first because it was the first to be used by the human race; secondly because the whole world uses it, though with different inflections and vocabularies; and finally because it is natural to us, while grammar exists more as an artefact.

(I. i. 2–4)

Latin, Dante is saying, and with it Greek and other languages of classical origin, are second-order languages. They are made up and have to be studied. The vernacular by contrast comes naturally. It is rooted in the basic structures of awareness, and is learnt—as Dante discovered from a famous chapter towards the beginning of Augustine's *Confessions* (I. viii)[4]—at our mother's knee. And with this the new and essentially anthropological-linguistic, as distinct from literary-philosophical, perspective of the *De vulgari* is established. Dante has gone beneath the literary-philosophical to affirm the nature of language as a feature of consciousness, as (again) the *form* in this or that individual or group of individuals of historical self-awareness. Soon, it is true, he will rise again to the level of the literary and aesthetic, and there, sure enough, will be Latin ready and waiting to assert itself as prior and paradigmatic. For the moment, however, and with an astonishing degree of socio-linguistic insight, he has established an alternative basis for setting up and resolving the question of language.

With this, the search for an Italian vernacular eloquence gets under way. Dante's perspective is long. He starts in the Garden of Eden, where man was endowed with a language lexically, morphologically, and syntactically, ready-formed—the language by means of which he was at once able to speak the name of God (I. iv. 4). Foolishly, however, he rose up in disobedience, first in Eden, then at the time of the Flood, and then at the time of Babel, when, insanely, his thoughts turned, not to equalling, but to surpassing his Maker in glory (I. vii. 4).[5] The result was chaos. Fragmented linguistically, and as a result morally, socially, and psychologically, men were destined henceforth to live in misunderstanding and mistrust, language from this point on pursuing its ungraced way of discrete and random evolution. Thus the first phase in the history of language, the phase beginning in Eden and ending at Babel, coincided with, and was dominated by, the notion of disgrace, of man's falling from favour with God. Now, however, the argument takes a curious turn, for the way of discrete and random evolution turns out in the post-Babelic phase to be the normal way of things, the proper and routine pattern. Disgrace is overtaken by development, Augustinian unnaturalism by Aristotelian naturalism:

No effect, then, we say, inasmuch as it is an effect, is more powerful than its cause, since nothing can bring about what it itself is not. And since, after the confusion which represented nothing but a forgetting of the earlier language, every language of ours (except the one first created by God) has been restored at our pleasure, and since man is a most unstable and changeable creature, his language cannot be lasting and constant, but rather, like all our other customs and fashions, must change according to time and place.

(I. ix. 6)

Prominent still is the notion of Fall and catastrophe, of the confusion or 'forgetting' which followed in the wake of Babel; but gradually emerging as the passage goes on is a sense of historical and geographical relativity. Everything comes and goes, and language is no exception. This in fact, taking his cue from the famous passage in Horace about words being like the autumn leaves (*Ars poetica* 60–72),[6] Dante had already stressed in the *Convivio*:

Onde vedemo ne le cittadi d'Italia, se bene volemo agguardare, da cinquanta anni in qua molti vocabuli essere spenti e nati e variati; onde se 'l picciol tempo così transmuta, molto più transmuta lo maggiore. Sì ch'io dico che, se coloro che partiron d'esta vita già sono mille anni tornassero a le loro cittadi, crederebbero la loro cittade essere occupata da gente strana, per la lingua da loro discordante. Di questo si parlerà altrove più compiutamente in uno libello ch'io intendo di fare, Dio concedente, di Volgare Eloquenza.

(*Conv.* I. v. 9–10; cf. *DVE* I. ix. 7)

But now it becomes the point of departure for a detailed history of the development of the vernacular from its post-Babelic fragmentation down to the contemporary Romance languages and to the linguistic diversity of Italy in particular. The major linguistic zones in Europe, determined, Dante explains (I. viii), by the pattern of migration after Babel, are three: the Slavic, Teutonic, and Saxon region to the north; the Greek region to the south east in the Balkan peninsula, the Aegean isles, and Asia Minor; and the Romance area to the south, including France, Provence, the Iberian peninsula, Italy, and Sicily. Each of these linguistic groups subsequently divided further, each subdivision, however, retaining something of the parent idiom. In the case of the Romance world the subdivision was again threefold, yielding the languages of *oc* (Provençal), *oïl* (French), and *sì* (Italian), all of which, Dante says, manifest their common origin lexically: 'Proof that these vernaculars derive from one and the same idiom is readily to hand in that they use the same words for many things, such as "God", "heaven", "love", "sea", "earth", "is", "lives", "dies", "loves", and almost everything else' (I. viii. 5). The prospect, therefore, is one of ceaseless linguistic diversification, observable, Dante says, even in a small city the size of Bologna, where speech habits vary from *borgo* to *borgo*, from street to street (I. ix. 4). Indeed, he goes on, it was in this context of change and decay that the 'art of grammar' ('gramatice facultatis') was founded as a sort of universal linguistic standard, a common currency of exchange over time and space:

> This is what inspired the inventors of the art of grammar, which is nothing but a particular, constant, and unchangeable usage unaffected by time and place. Since this was settled by the common consent of many peoples, it appears free from any one individual initiative and thus incapable of variation. They invented it lest, owing to the variation of speech fluctuating according to the will of this or that individual, we should fail either wholly or in part to have knowledge of the thoughts and deeds of the ancients or of those whom distance makes strange to us.
>
> (I. ix. 11)[7]

Who these *inventores* or founders of grammar are, and how they relate to the *positores* or legislators mentioned at the beginning of the next chapter, is not clear. It is, perhaps, a matter of the difference between speculative and practical grammar, between those who excogitate the ground rules, and those who elaborate them in respect of this or that particular language. But this, though it bears a number of intriguing implications for Dante's view of the relationship between Latin and Italian—for his view of the way in which 'speculatively', as distinct from culturally, Latin might be said to presuppose the prior categories of the vernacular—is a parenthesis, Dante's main concern being with linguistic relativity and with the consequences of this for vernacular eloquence. The problem, then, is to define amid all this variety a form of language properly representative of the Italian-speaking world in particular, something Dante sets out to do empirically, by reviewing the various contenders—there are at least fourteen (I. x. 7)—and by weeding out the weakest. In fact, his procedure is only apparently empirical, for its leading criteria—those of lyrical sweetness and of a (now cultural) closeness to Latin in point of grammar—at once emerge in I. x. 2:

The third too, the language of the Italians, claims pre-eminence on two special grounds: first, because those who have written the sweetest and most subtle poetry in the vernacular have been of its household and family, as were Cino da Pistoia and his friend; and secondly, because they seem to rely more on the grammar of Latin, which is the common basis of all three—an argument which, looked at rationally, is of great moment.

Straightaway, therefore, the anthropologically 'nobler' vernacular is—as *written*, as taken up and used by such as Cino and his friend for the purposes of lyric poetry—brought into dependence on Latin, with, on the technical front (in respect of sweetness and subtlety) a distinct leaning in the direction of contemporary literary Tuscan. For the other spoken forms of Italian, and even as we shall see for Tuscan itself as *spoken*, all this spells the end as far as their candidature for the *vulgare illustre* is concerned. So, for example, the phonologically and orthographically aberrant *messure* of Roman (the Tuscan *messere*), the morphologically misconceived *enter* of Milanese and Bergamasque (the Latin *intra*), the brusque accentuation (*accentus enormitas*) of the Aquileian *Ces fas-tu?* (the Tuscan *Che fai tu?*), the inept Latinism (i.e. mistaken agreement) of the Sardinian *dominus nova* and *domus novus*, the structural and consonantal irregularity of the Apulian *bòlzera* and *chiangesse* (the Tuscan *vorrei* and *piangesse*), the blandness of the Romagnolo *oclo meo* and *corada mea* (the Tuscan *occhio mio* and *cuor mio*), the abrasive Genoese fricative *z*, the 'rude asperity' ('rudis asperitas'—Dante has in mind the consequences of syncopation) of the Brescian and Paduan *mercò* and *bontè* (the Latin *mercatus*, *bonitas*), and the coarse regionalism of the Venetian *plaghe* and *verras* (the Tuscan *piaghe* and *verrai*)—all of these Dante finds technically and acoustically offensive.

But even Tuscan, Bolognese, and Sicilian, the most likely candidates for the *illustre*, will not do; for whatever of excellence has come about in, say, Tuscan—and Dante has in mind principally the *stilnovisti* Guido Cavalcanti, Lapo Gianni, Cino, and 'one other', 'unum alium'—has come about as a result, not of espousing, but of eschewing the language as spoken (I. xiii. 5). The language as spoken is full of localisms, of barbarisms, and of solecisms. Take, for example, the Florentines. They say 'Manichiamo (for *mangiamo*) introcque (for *intanto*) che noi non facciamo altro' ('Let us eat betimes—it's all we ever do')—nothing if not vulgar. And the same goes for the inhabitants of Lucca, Siena, Arezzo, and Pisa, all of them guilty of the most deplorable municipalism ('non curialia sed municipalia tantum invenientur', I. xiii. 1). Empirically, then, or by means simply of looking round, there is no solution; Tuscan, Bolognese, and Sicilian, they none of them measure up, as spoken, to the exacting standards of the *illustre*.

So now Dante tries another tack, a reasonable rather than an empirical line of argument ('rationabilius investigemus de illa', I. xvi. 1). Everything, he says, inasmuch as it belongs to this or that species or family, has a 'simple sign' or *signum simplicissimum* against which it can be measured and evaluated.[8] Numbers, for instance, are measured against unity. They are 'called greater or lesser according to their distance from, or proximity to, one' (xvi. 2). Colours are judged against white. The more white they have, the more visible they are. Activities too are measured in the same way. As moral beings—as beings of reasonable deciding and doing—we have virtue or the *honestum* as the standard by which we are judged. As citizens, we have the law. As Italians, we have the 'simple standards of custom, manner, and speech by which the deeds of Italians are weighed and measured' (xvi. 3). And here, Dante says, lies the solution to the problem in hand, a way of defining the *vulgare illustre*; for the highest standards common to Italians as a whole, the standards of custom, habit, and speech, include too the highest *linguistic* standard, the 'signum simplicissimum' of speech. The *vulgare illustre*, then, is the 'illustrious, cardinal, courtly, and curial' vernacular against which all the spoken vernaculars stand to be 'measured, weighed, and compared': 'dicimus illustre, cardinale, aulicum et curiale vulgare in Latio . . . quo municipalia vulgaria omnia Latinorum mensurantur et ponderantur et comparantur' (xvi. 6).

Here, though, Dante's argument becomes peculiarly fraught, painfully alive, once more, to the notion of exile. The description itself, the notion of the *vulgare illustre* as illustrious, cardinal, courtly, and curial, is unproblematic. The 'signum simplicissimum' of Italian vernacular eloquence is illustrious in that it lights up morally and intellectually both the user and the listener. It has the power to illumine inwardly, to glorify through its resplendence (xvii. 2). It is cardinal in that, like a hinge (*cardo*), it is a point about which the whole vernacular turns (xviii. 1).[9] It is courtly in that, as the common property of the Italian people, it resides in the highest court of the land, representing it among those who are strangers and visitors there (xviii. 3). And it is curial in that, like a rule or balance, it is normative (xviii. 4). But—and this is the point—the *vulgare illustre*, like Dante, is in exile. It has no home. It is scattered about the country in the hearts and minds of those who, eschewing the solecisms and barbarisms of the spoken tongue, have fashioned it despite rather than because of their circumstances. The idea appears first in Chapter xii, in the course of Dante's review of the spoken language in general and of Sicilian in particular. Sicilian, like Bolognese and Florentine, lays claim to literary excellence. But literary excellence—and here lies one of the main elements in the inspiration of this first book of the *De vulgari eloquentia*—is bound up in his mind with political excellence, with the notion of government by men of moral and intellectual virtue. Thus the first home of the *vulgare illustre* was the court of the Hohenstaufen in Sicily, the court of Frederick II and of his son Manfred, who, 'scorning to live like beasts', set an example of sane, stable, and civilized princely rule:

And first let us examine the genius of Sicilian, for this above all others seems to demand recognition, both because whatever poetry we Italians have written is called Sicilian, and because we find that many learned natives of Sicily have written serious verse, such as the canzoni *Ancor che l'aigua per lo foco lassi* and *Amor, che lungiamente m'hai menato*.

But this fame of Trinacria, considered in respect of its real significance, appears to have survived solely as a reproach to the Italian princes, who, far from heroes, live in their pride like commoners. For, as long as fortune permitted, those great heroes the Emperor Frederick and his nobly-born son Manfred, showing forth the nobility and righteousness of their souls, lived like men, scorning the bestial. And this is why those of noble spirit and gracious endowment sought to be one with the majesty of such great princes, so that whatever the finest Italian spirits achieved in literature first saw the light in the age and at the court of these sovereigns. And because Sicily was the royal seat, it came about that whatever our predecessors wrote in the vernacular was called Sicilian—a usage to which we hold fast and which even our successors will be unable to change.

Racha! Racha! How now sounds the new Frederick's trumpet, the cymbal of Charles II, the horns of the powerful marquises Giovanni and Azzo, and the pipes of all the other great lords? Nothing but 'Hither, murderers, traitors, and avaricious!'

(I. xii. 2–5)

With the Hohenstaufen, the *vulgare illustre* found a home. Its nobility was that of the court itself, a reflection of the court's accomplishment in every aspect of art and science. Its majesty was that of its great patrons, and its integrity formal confirmation of their just rule and wise counsels. And now? Double catastrophe. With the emperors long gone and Italy abandoned to the petty tyrant, the *illustre* is in exile. Destitute, humiliated, and a stranger in its own land, it wanders from one place of refuge to another. True, the occasional poet from Bologna, Florence, and Padua has done it proud, but by and large it is an outcast, a forlorn seeker after peace and security: 'That is why all who frequent royal courts speak the illustrious vernacular; and why our own *illustre* wanders from place to place like a foreigner, seeking hospitality where it can in humble dwellings, for we have no court' (I. xviii. 3). Hence the ultimately tragic tone of this first book of the *De vulgari eloquentia*, its sense, not of triumph and acclaim, but of estrangement and indifference.

Dante's solution, then, to the problem of Italian vernacular eloquence is a literary one. For all his intent to proceed anthropologically, historically, geographically, anything in fact other than literarily, he is forced bit by bit into associating the *vulgare illustre* with the great 'Sicilian' poets by whom it has from time to time been realized; hence the nature of the second book as an *ars poetica*, as a manual of poetic theory.

The tradition of poetic manuals, or *poetriae*, goes back to the great rhetorical treatises of antiquity, to those of Aristotle, of Cicero and the pseudo-Ciceronian *Rhetorica ad Herennium*, of Horace of the *Ars poetica*, and of Quintilian of the *Institutio oratoria*. These in turn were taken up and developed in the latter part of the Middle Ages by authors such as Matthew of Vendôme in his *Ars versificatoria* (*c.* 1175), Geoffrey of Vinsauf in his *Poetria nova* (c. 1210), John of Garland in his *Parisiana poetria* (completed c. 1235), and, closer to home, by Brunetto Latini in the third book of his *Tresor* (c. 1260). Some at least of these texts—Latini's certainly—Dante must have known well, for early on in the *De vulgari* (I. i. 1) he speaks of the 'sweet mixture' or 'hydromel' of his own and of others' inspiration in the treatise, a sweet mixture most obviously discernible in the Ciceronian aspects of the second book. But there the resemblance ends, for where the manuals are interested in the analysis of form in and for itself, or at most in the analysis of form inasmuch as it relates to this or that particular theme—the whole question thus remaining external to conscience, a matter merely of good taste and of technical decorum—what concerns Dante is the function of form as an index of spirituality in the poet himself. The whole question, in other words, is internalized, rooted in the intimate movement of thought and feeling. Now the *stilnovisti*, it is true, had come a good way along this road. Impatient of the technically and imaginatively otiose, they too had wished to link poetic form with the movement of thought and feeling, with the 'dittar dentro' within. And this, in its time, had produced deep changes in the theory and practice of 'courtly' lyric verse. But here in the *De vulgari eloquentia* the youthful intuitions of the *stilnovisti* are taken up in the altogether more 'virile' context (*Conv.* I. i. 16) of moral and intellectual becoming generally, of mature self-possession in and through the word.

First, though, the formalities: what, for Dante, is poetry?

Poetry, Dante says, in the key formula of II. iv. 2, is 'nothing other than invention fashioned with music and rhetoric': 'nichil aliud est quam fictio rethorica musicaque poita.' On the one hand, there is the movement of mind and imagination, while on the other there is the musical and rhetorical strategy whereby this is given intelligible form. But in what sense is the poem musical? Dante has two things in mind, the first musical in a sense approaching the modern usage of the term, and the second more 'literary-musical'. The first sense, the 'musical' sense proper, looks to the tradition of practical music-making in France and Provence. Here we are fortunate in having to hand a good number of extant *chansonniers*, of collections of words and music dating back to the thirteenth century. These witness to a developed but austere sense of musical structure.[10] The various parts of the song must preserve a perfect balance and proportion. Metrically especially, they must be symmetrical. And this passes over into Dante's *ars cantionis*, likewise envisaging a susceptibility on the part of the poem to musical setting ('every stanza must be harmonized in such a way as to be set to music', II. x. 2). The rules, then, are these. First comes the *cantus divisio* or primary division of the stanza—the fundamental poetic unit—into two (there being usually two melodies in the song): the *frons* and the *sirima* (or *sirma*). The *frons* in turn, and occasionally the *sirima*, subdivides into two equal feet or *pedes* (*versus* or *voltae* in the case of the *sirima*) to allow for the repetition of the first melody. The *frons* and the *sirima* are joined by a 'key line' or *verso chiave* belonging syntactically to the *sirima* but rhyming with the last line of the *frons*. Thus the whole thing looks like this:

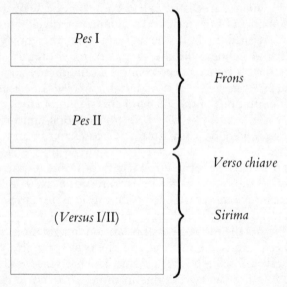

Next comes the *partium habitudo* or relative size and constitution of these parts. Again the rules are simple, the minimum necessary to ensure overall symmetry: the *pedes*, and, where they exist, the *versus*, must be of similar length and constitution. The poet, in other words, and again for reasons of musical setting, must preserve throughout a similar distribution of lines and of line-types (mainly hendecasyllables and *settenari*). Also, he must pay particular attention to rhyme, for Dante a matter of the greatest aesthetic interest ('on this especially depends the sweetness of harmony in the whole', II. xiii. 4). Rhyme, whether at the end of a line or in the middle, is a means of securing consistency and variety in the stanza, variety especially—the contrasting effect of the 'smooth mixed in with the harsh' (II. xiii. 13)—commanding Dante's admiration. Also, it is a means of confirming finality in the stanza, of bringing it to a peaceful conclusion: 'particularly beautiful, however, are those final-line inflections which, rhymed, fall together into silence' (II. xiii. 8).

The third and final aspect of this idea of the poem as song—so far we have had the *cantus divisio* or dividing up of the stanza into its principal parts, and the *partium habitudo* or relative constitution of those parts—is the number and type of lines in each part, the *numerus carminum et sillabarum*. Here Dante gets no further than some preliminary observations on subject matter. Sometimes, he says (II. xiv. 2), we speak persuasively or dissuasively, sometimes in praise, sometimes ironically, sometimes judgementally, and so on. The size of the stanza, therefore, and of its constituent parts, depends on its mood and purpose. Let those, Dante concludes, who propose to speak harshly in their verse make short work of it; as for the rest, let them proceed more fulsomely, with which, for reasons we shall come to shortly, the book all of a sudden fades away.

But the poem is 'musical' too in its sense of verbal harmony, in its sense of the way in which different sounds enmesh to create a pleasing whole. Here Dante takes a decidedly independent line with respect to his predecessors, to Matthew of Vendôme, for example, or to Geoffrey of Vinsauf. They had insisted on expunging everything vaguely agrarian (as Geoffrey of Vinsauf puts it), everything liable to offend the ear in its harshness: 'Thirdly, let your choice of words be civil rather than rustic' (*Poetria nova* 83–4). Matthew of Vendôme was even firmer. 'There are certain ragged words', he says in the *Ars versificatoria* (ii. 46), 'that ought never to appear in a metrical composition, words which, having been so to speak anathematized, are utterly unfit to hold company with others. Among them are such words as *porro* (further), *autem* (however), *quoque* (also), and other expressions of a similar syntactic or associative function. These words, since they detract from the grace of the poem, ought completely to be eliminated from poetry.' The ideal, clearly, is exclusive. Some sounds are simply not welcome. But Dante's ideal is inclusive, based on a sense of the way in which different sounds—the 'shaggy' and the 'groomed' as he puts it (the 'yrsutus' and the 'pexus')—interact with one another to produce a balanced and differentiated whole: 'Ornamental, by contrast, is the term we give to all those polysyllabic words which, when mixed with "groomed" words, produce a pleasantly and harmoniously constructed whole' (II. vii. 6). Gone, therefore, is the bland uniformity of the rhetorical period as traditionally conceived, its avoidance of the brisk and assertive. The poem is more accommodating, acoustically more lively.

With this we come to the rhetorical side of Dante's formula ('poesi[s] . . . nichil aliud est quam fictio *rethorica* musicaque poita'), to the 'sweetness' bestowed on discourse by the disciplined use of rhetorical devices. The idea is Ciceronian and is celebrated by Dante in the *Convivio*. Rhetoric, he says in *Convivio* II. xiii. 14, is the sweetest of the sciences, sweetness being its main business: 'ché la Rettorica è soavissima di tutte le altre scienze, però che a ciò principalmente intende . . .' If, then, its beauty lies in order, in the regular structuring of the parts in relation to the whole (ibid. II. xi. 9), its function lies in winning over by per-suasion, by the 'making sweet' of the prose or poetic line (ibid. II. vi. 6). There are, more precisely, two points at which Dante is in touch with his rhetorical tradition: (*a*) in respect of the *genera dicendi* or levels of style, and (*b*) in respect of levels of construction (*constructio*). The key text as far as (*a*) is concerned is the pseudo-Ciceronian *Rhetorica ad Herennium*, which distinguishes three levels of diction, the grave, the middling, and the simple (*gravis*, *mediocris*, and *extenuata*, IV. viii. 11), terms denoting degrees of syntactical and lexical refinement ranging from the ornate and impressive to the colloquial and commonplace:

There are, then, three kinds of style, called types, to which discourse, if faultless, confines itself. The first we call the Grand; the second, the Middle; and the third, the Simple. The Grand consists of a smooth and ornate arrangement of impressive words. The Middle type consists of words of a lower, yet not of the lowest and most colloquial, class of words. The Simple type is brought down even to the most current idiom of standard speech.

Similarly, Isidore of Seville, in the *Etymologies* (II. xvii), distinguishes between the grand (*grandiloquum*), the middling (*medium*), and the humble (*humilis*), linking as he does so the various levels of diction with the dignity of the subject-matter:

These, then, are the three levels of diction: the humble, the middling, and the grand. When, therefore, the matter to hand is exalted, the grand style is properly to be chosen for its utterance; when the matter to hand is trivial, the plain and unadorned style; and when it is in between, the moderate style. In trivial matters nothing grandiloquent or sublime is to be said. Rather, the manner of speech should be mild and down to earth. In greater matters, though, when we are speaking of God, say, or of man's well-being, then greater magnificence and splendour is to be displayed. In middling matters, where nothing is at stake save the hearer's entertainment, a manner of speaking between the two is what is needed.[11]

Dante too distinguishes three levels of style—the tragic, the comic, and the elegiac—related to, and determined by, subject-matter. Noble matters—matters, as we shall see, of *salus*, *venus*, and *virtus*, of well-being, love, and virtue—demand the high or 'tragic' style, while others are properly dealt with either in the comic style, a vigorous, 'realistic' style, or in the elegiac style, a plaintive style especially appropriate, according to John of Garland (*Parisiana poetria* v. 365–72), to the lamenting of lost or crossed loves ('dolores amancium'):

Next we must be able to determine, among those things that present themselves as possible subjects for poetry, whether a tragic, comic, or elegiac treatment is appropriate; if tragic, then we introduce a loftier style, and if comic we introduce a lowlier style, while by elegiac we mean a base style. If our song appears to demand the tragic style, then we must use the *vulgare illustre*, and in consequence bind our words in the form of a canzone. If the comic style is called for, then we must use sometimes the lowest and sometimes the middle form of the vernacular, how to choose between them being a matter for demonstration in Book IV. If the elegiac is called for, then only the lowest form of the vernacular is to be used.

(*DVE* II. iv. 5–6)

So far, then, we have a distinction of literary styles or levels of diction (*genera dicendi*) much along the lines of what we find in Cicero, in Quintilian, and in the neo-Ciceronian rhetoricians of the late Middle Ages, the terms 'tragedy' and 'comedy' referring primarily to style or register, to the *how* rather than to the *what* of the poet's discourse.[12]

More important, however, than the *genera dicendi*—for Dante is concerned in this second book of the *De vulgari* only with the high style, or *stilus tragicus*—are the appropriate levels of 'construction' (*constructio*). By 'construction' Dante means the rules governing the arrangement of words (II. vi. 2), which means that—the *choice* of words apart (the poet, Dante says in II. vii. 5, should choose only those words which leave a 'sweet residue of sound and sense in the soul')—the question is predominantly a syntactical one. And here Dante is at his most spirited. There are, he says, if we include the level of elementary grammatical propriety (*constructio congrua*), four grades of construction (*gradus constructionum*). First comes the ordinary subject-verb-object arrangement of words within the sentence: 'Petrus amat multum dominam Bertam' ('Peter loves very much lady Bertha'). No problem here; everything is plain and unadorned. Next, however, and representing the first degree of rhetorical elaboration, come periods of the type 'Piget me cunctis pietate maiorem, quicunque in exilio tabescentes patriam tantum sompniando revisunt' ('I, surpassing others in compassion, grieve for all those who, wasting away in the langour of exile, see the fatherland only in dreams'), where much in evidence is the artfulness of the rhetorician. To begin with, there is the artificial order (*ordo artificialis*) of the constituent clauses—the initial *prolepsis* or subject-verb inversion of the opening phrase. Then there is the separating out of the *me* and its qualifying *maiorem*. And finally, though invisible to the modern reader until pointed out, there is the rhythmic cadencing (*cursus*) of the period in keeping with the principles of the *ars dictaminis*.[13] All very sophisticated, but even so, Dante says (II. vi. 4), no more than might be expected of scholars and schoolmasters. Altogether grander is the following: 'Laudabilis discretio marchionis Estensis, et sua magnificentia preparata, cunctis illum facit esse dilectum' ('Praiseworthy is the discerning of the Marquis of Este and ready is his bounteousness, making him beloved of all'). Now the Marquis of Este (Azzo VIII of Este) we have already met in I. xii. 5, where he is noted for his avarice and ignobility—a line taken too in the *Commedia* (*Inf.* XII. 111–12; *Purg.* v. 77). Thus Dante's period here in II.vi. 4 takes on new force. The familiar features of 'easy ornamentation' (*ornatus facilis*)—inversion, separation, repetition, and the rest—combine with the difficult figure of irony (*permutatio* in the *Ad Herennium*) to produce a statement of unusually stunning power. But even that is modest; for over and above the 'sapidus et venustus' (the tasteful and comely) comes the 'sapidus et venustus etiam et excelsus' (the tasteful, the comely, and the exalted). Dante's example—rooted, we must note again, in the substance of his own experience—runs as follows: 'Eiecta maxima parte florum de sinu tuo, Florentia, nequicquam Trinacriam Totila secundus adivit', which means 'Having cast out, Florence, the noblest of your flowers, vain was the second Totila's descent into Italy' (where the second Totila is Charles of Valois, who, having descended into Italy in 1301 with a view to pacifying Tuscany in particular, succeeded merely in making matters worse). Technique here is consummate. Inversion and the *cursus* are accompanied in the first clause (absolute in construction) by *abbreviatio*, and in the rest by apostrophe, personification, hyperbole, *antonomasia* (or *pronominatio*—the 'Totila secundus'), and, above all, by metaphor, here as before (in the *petrose*) the jewel in the crown.[14] With this, Dante says (and he gives a range of further examples from the Provençal and Italian poets), we have reached the apogee of rhetorical art, the high plains of tragic construction. Never mind Guittone d'Arezzo and his like. They, Dante says, are inept municipals, unfit even to voice their opinions in this matter. Here instead we are in the great tradition of the ancients, of those who, like Virgil, Ovid, Statius, and Lucan, and (on the prose side) Livy, Pliny, Frontinus, and Orosius, were true masters of rhetorical art:

You must not be surprised, reader, that we call to mind so many authors; for we can show only by examples of this kind what we mean by the best construction. It would, perhaps, be useful, for the purposes of acquiring the habit of such construction, if we had examined the 'regular' poets, Virgil, Ovid in his *Metamorphoses*, Statius and Lucan, and others who have used the finest prose, such as Titus Livy, Pliny,

Frontinus, and Paul Orosius, and many others whom loving solicitude causes us to frequent. Let, then, those followers of ignorance who extol Guittone d'Arezzo and others who have never shaken off the habit of using plebeian words and constructions fall silent.

(II. vi. 7–8)

And with this we return to the leading idea of this second book of the *De vulgari eloquentia*, to the notion of form as 'an index of spirituality in the poet himself'; for underlying and quickening these technical chapters of the book is a sense of form's responsiveness, not simply to the matter in hand, but to the quality of the author's own being, to the substance of what the author himself actually is as a creature of reason and of moral self-determination. The key chapters now are i–iv, and again the argument is worth following step by step. The first thing to recognize, Dante says, taking up where he had left off in Book I, is that, as far as Italian is concerned, poetry rather than prose furnishes the linguistic model. Prose writers merely follow in the footsteps of the poets, of those who 'bind their discourse musically' (II. i. 1). But this raises questions about who should use the *illustre*, and about appropriate themes and forms (sonnets, ballads, canzoni, etc.). Not all poets, Dante says in Chapter i, should use the *illustre*, but only those of native wit and learning ('ingenium et scientia'); for the best deserves the best, and the *illustre*, the *signum simplicissimum* of the Italian vernacular, deserves as its representatives only those most accomplished in the ways of art and science. Again, not all themes are appropriate for use with the *illustre*, but only those relating to the nature of man as a threefold creature of vegetative, animal, and rational constitution dwelling in time and space: the themes of *salus* (well-being), *venus* (love), and *virtus* (virtue). And not all forms are appropriate for use with the *illustre*, but only those equal to, and traditionally associated with, man's noblest thoughts and feelings—only the canzone (Chapter iii). Dante's drift is plain. Impatient with form for form's sake, with the irresponsible use of form either by the unqualified or in unsuitable circumstances, he unfolds the question in terms of the properties of personality, of human nature as determined in this or that individual. True, the technical aspect remains prominent. Poets, Dante insists (II. iv.

1-3), are craftsmen. They have an art to learn and an apprenticeship to serve. Let no one, he says, be under any illusion about that. For far too long vernacular lyric art has been conducted casually, 'casualiter', without proper discipline. But technique serves the purposes of confession, with the result that not only must the poet know his craft, but (*a*) he must adjust the burden of his discourse to the strength of his shoulders,[15] and (*b*) he must drink deeply of Helicon and immerse himself in the ways of art and science. Then, and only then, may he take up his plectrum. The way is hard, Dante says, and the demands made upon the poet by his calling endless ('hic opus et labor est').[16] But the rewards are stupendous; for the poet who thus knows himself, who thus lays hold of his inmost thoughts and feelings, and who thus bends the word to the exigencies of conscience, *is*, as a man and as a poet, completely and triumphantly. His being is consummate, fully transparent in respect of its intimate reasons and overall purposes. He is in short, blessed of the gods, a high-soaring, ecstatic spirit:

Let everyone, then, beware and understand what we are saying, which is that anyone intending to sing of these three subjects, either in themselves or in respect of those things flowing directly or simply from them, must first drink of Helicon and tune every string to perfection, only then taking up the plectrum with confidence. But the real hard work lies in the proper exercise of caution and judgement, something that can never be achieved without constant mental effort and perseverance in the ways of art and science. These are those whom the poet of *Aeneid* VI (though he is speaking figuratively) calls beloved—nay sons—of the gods, raised to the skies by their glowing excellence. Let the folly of those, therefore, who, untouched by art and science and trusting solely in their native wit, rush in to sing of the highest things in the highest style, be confounded. Let them desist from such presumption. And if they are geese by nature or by idleness, let them not wish to imitate the eagle that seeks the stars.

(II. iv. 9–11)

And with this, Dante is home; for having linked the practice of verse-making with the verification in the poet of his essential humanity, he has confirmed once and for all the anti-formalist thrust of his literary aesthetic. Old habits of mind—represented in the way we have seen by the definition of procedures—linger on, it is true. But the whole question is subject now to moral intensification. Poetry, far from marginal and dispensable (as in the case, say, of a Dante da Maiano or of a Forese Donati), stands and functions at the centre of personality, at the point of maximum self-engagement and concern. Hence the seriousness, the *serietà terribile*, with which Dante tackles the whole issue; lexis, syntax, number, rhythm, structure, in short the whole object of art's 'assiduity' (*assiduitas artis*), reflects the truth of the poet's moral and intellectual being. It is but one aspect of—as one modern critic puts it[17]—his 'ansia di perfezione', of his search through form for moral perfection.

What, then, can we say about the *De vulgari eloquentia* by way of conclusion? Two things: (*a*) that as an essay in the theory of language and of literature it has about it a peculiar impossibility; and (*b*) that its impossibility renders it one of the most precious documents we have from the hand of a great European poet.

First, therefore, its impossibility, its—to begin with—linguistic impossibility. The *De vulgari* starts out, as we have seen, as a treatise on language. It sets out in the wake of the *Convivio* to define a standard of Italian vernacular eloquence. But casting around him for a model or set of principles on which to base his argument, Dante is forced to look for it among the poets, and among poets, moreover, representative of the high or tragic style—the Sicilians and *stilnovisti*. The result of this is an ideal not so much reflecting as repudiating the living language. Inclusivity gives way to selectivity, to a defining of Italian in terms of a way of thinking and writing self-consciously particular, cliquish even, in character. True, the middle and low styles were to have been treated in Books III and IV (II. iv. 6). But by then it would have been too late; for the *illustre*, inasmuch as it had been realized at all, had been realized only by the select few resolutely opposed to poeticizing in the spoken language, which, therefore, virtually by definition, emerges as something *less* than Italian, as something merely municipal. The inadequacy of this as a basis for the kind of enterprise represented by the *Commedia* hardly needs stressing. Bound by its leading equation of the *vulgare illustre* and the *stilus tragicus*, the *De vulgari eloquentia* is doomed to partiality and incompleteness. It is crippled by its own solution.

There is a different kind of impossibility on the aesthetic front. Dante's thoughts about poetry and about beauty in poetry have, as we have seen, to be considered against the background of the neo-Ciceronian rhetoric of the manuals, of the kind of technical guides to composition represented by the *Ars versificatoria* or by the *Parisiana poetria* or *Poetria nova*. Beauty here is a matter, pre-eminently, of addition (*additio*) or superimposition. There, duly catalogued, are the myriad flowers of thought and form ready to be plucked and woven into a 'festive' garland, as Matthew of Vendôme puts it,[18] of poetic discourse. Certainly, tact is called for. All the rhetoricians are agreed about that. The outward display has to reflect the splendour of thought and feeling. But that does not alter the basic sense of what poetry is about; for it is still a matter of imposing the formal strategies of art, of applying the mechanisms currently and traditionally available to the professional wordsmith. This is what we mean by 'formalism' or 'aestheticism'—the identification and separating out of an aesthetic moment proper, the notion of aesthetic intervention. Nothing, however, could be further from Dante's mind in the *De vulgari eloquentia*, for poetic form, far from being 'added on' to the substance of the poem, is part of that substance. It is substance under the aspect of intelligibility. But—and this is the point—the aesthetic question thus conceived defies prescription. It defies prescription for the simple reason that it is no longer possible to say in advance what will or will not be aesthetically acceptable. All that can be done is to ponder the interaction in the poem itself of form and substance—which is why Dante goes out of his way to apologize for his long-windedness: 'Don't be surprised', he says at II. vi. 7, 'if I am having to recall to mind so many instances. Some things, like the question of *constructio*, cannot be handled otherwise. The only way to appreciate what is right is to see it in operation.' Example is the only workable strategy. Given the wealth of formal devices available to the would-be poet in the high style, those which he at last chooses, and the excellence, therefore, of his final utterance, all depend on how far the word responds *in casu*, in this or that instance, to the inner and otherwise unsearchable 'dittar dentro'. And for that there is no legislating. Hence the sad, but not entirely premature, petering out of Dante's book in mid sentence, in another moment of 'all-depending'. Aesthetically no less than linguistically it is doomed to incompletion. It does not work because it cannot work.

But here as elsewhere in Dante impossibility testifies to

grandeur, to yet a further attempt to make plain the principles of his activity so far. The project is typically Dantean; never content to allow these principles to remain shrouded in uncertainty or vague in statement, he cannot rest until they are brought into the light of day, into the realm of categorical certainty. First, then, there is the question of language. Confronted by the fact of a language as yet less than mature philosophically and expressively, Dante, moved alternately by love and by pride, undertakes to remedy the situation, to fashion (in the *Convivio*) and to define (in the *De vulgari*) a new form of vernacular eloquence. Nowhere more than here—with the possible exception of the 'Io non Enëa, io non Paulo sono' clause of *Inferno* II. 32—is that recognizably Dantean sense of his own historical indispensability more marked. And what goes for language goes for literature too. Heir to a predominantly aestheticizing rhetorical sensibility, he sets out to promote a sense of style, not merely as 'convenience' or suitability, but as co-efficient or co-creator, as a principle in this or that individual of his specifically human *being*. In this sense, the *De vulgari eloquentia*, impossible as it is as a *poetria* or treatise *de arte scribendi*, looks forward to the triumph of the *Commedia*, the supreme affirmation of existential totality in and through the word.

4
The *Monarchia*, the Letters, the *De situ*, and the Eclogues

(i)

The Monarchia: *introduction and chronology*

LIKE the *Vita nuova*, the *Monarchia* has both a public and a private face. Publicly, it forms part of the political controversy culminating in the conflict at the turn of the thirteenth century between Boniface VIII and the French king Philip IV (Philip the Fair) and continuing in the struggle for temporal dominion under Clement V and John XXII.[1] Thus Dante presses the claims of the Holy Roman Empire over and against those of the papacy, *both*, in his view, deriving immediately from God. Privately, the *Monarchia* takes up, develops, and indeed exhausts, the 'humanist' emphases of his philosophical career so far, the separating out, again in favour of the here and now, of the various orders of activity proper to man as man. The process began, as we have seen, in the *Convivio*. But where in the *Convivio* it had been pursued in a predominantly moral and active register, here instead, and far more dangerously, it is pursued in an intellectual or speculative register, which brought Dante and his book into dire disfavour with the Church. This, then, is a book of multiple interest: on the one hand, it is a political treatise of (especially as it goes on) consummate dialectical skill rooted in the concerns of the time; on the other, it is a monument to the Christian humanism characteristic, though not for the most part so dramatically, of classical Catholic theology generally.

But first, chronology, a question which in the case of the *Monarchia* bears on its interpretation. The early and late limits for its composition are 1308 or thereabouts and 1321, the year Dante died. In these years papal-imperial relations took a turn for the worse. In 1308, Henry VII of Luxemburg was elected emperor, and preparations began for his coronation in Rome.

The Pope, Clement V, troubled by Ghibelline dissidence in the north of Italy, looked favourably on the prospect of a fresh imperial descent, for whatever this might represent ideologically, by way of a return to a proper papal-imperial dualism, it would at the same time bring the Ghibellines to heel. Clement therefore wrote to Henry in July 1309 recognizing his election and suggesting a date for his coronation. The letter begins with an ample statement of the twofold character of human authority, of the complementary role of priests and princes:

The unfathomable wisdom of the Most High, in so disposing earthly things after the model of heavenly things and in so causing the waters to flow from the mountain heights that by the fruit of His works the world here below might prosperously be ordered, just as He appointed in the firmament of the heavens two chief luminaries to enlighten the world by turns, so on earth He made especial and supreme provision in the form of the priesthood and of the empire, for the full direction and governance of things spiritual and mundane . . .[2]

—a theme taken up in the encyclical *Exultet in gloria* of September 1310, and by Dante in his letter of about the same time to the princes and people of Italy: 'This is he whom Peter, the Vicar of God, exhorts us to honour, and whom Clement, the present successor of Peter, illumines with the light of Apostolic benediction, that where the spiritual ray suffices not, there the splendour of the lesser luminary may lend its light' (*Ep.* v. 10). Now if the *Monarchia* belongs to this period, then it must be regarded as a kind of manifesto, optimistic, constructive, and conciliatory. And, for all its polemical force, it *is* constructive and conciliatory. Clerical greed is passed off as 'zeal for the keys' (III. iii. 7), and imperial authority is acknowledged as standing in need of papal grace (III. iv. 20), all of which proceeds less from the polemicist than from the philosopher, less from the critic than from the advocate of a new way.

But by the time of Henry's death in August 1313 unity of purpose had crumbled. His coronation in Rome in June 1312 had been a humiliating affair. The city was alive with Guelph and Angevin opponents to the imperial cause, and Henry had to be crowned, not in St Peter's, but in St John Lateran, with the Bishop of Ostia standing in for the absent Clement. Foremost among the dissidents was the Pope's protégé Robert of Naples, whom Henry summoned to appear before him in Pisa, this in turn enraging the Pope and causing him to rethink the relationship of papal and imperial power; hence the new hierocracy of *Romani principes* and *Pastoralis cura*, the twin constitutions of March 1314.³ The emperor, Clement insists, is a vassal of the Church, and in the case of an imperial interregnum—as now, after Henry's death—secular authority reverts *de iure* to the pope, who then wields both swords directly. In fact, Clement died in 1314 and the whole matter passed into abeyance; but John XXII, elected in 1316 and faced with rival claimants to the empire (Frederick of Austria and Lewis of Bavaria), took it up with a vengeance, his bull *Si fratrum* (1317) yielding nothing to *Unam sanctam* (Boniface VIII's bull of 1302) in point of ecclesiastical absolutism. God, it said, through Peter, has committed all authority in heaven and on earth to the pope. All pretenders to the imperial throne must submit themselves for his approval.⁴

What happened next—as dramatic as it is undignified—is familiar enough. John XXII was offering an ultimatum: obedience or else excommunication. Those who refused to submit were deemed to be in a state of rebellion against the Church, and by 1320 rebellion was associated with heresy. Especially suspect were the Ghibellines, necromancers to a man, and their arch-patron Lewis of Bavaria, who in 1323 was indicted by the Pope as a *fautor hereticorum*, a harbourer of heretics.⁵ Lewis in turn accused the Pope of being a sower of discord, a traitor to God's purposes for men on earth, and a cause of scandal in Christendom,⁶ whereupon, especially with the emperor's espousal of the Franciscan cause, schism was inevitable. Ghibellines, Spirituals, and other malcontents in tow, he made his way to Rome, shortly to appoint in opposition to the 'priest John' the anti-pope Nicholas. A decade and more's papal intransigence had reaped its dreadful harvest.

By this time Dante had been dead six years. But the events leading up to schism had been in train for a good ten, and Dante, before he died, had pronounced vigorously on the rapaciousness of the popes:

> In vesta di pastor lupi rapaci
> si veggion di qua sù per tutti i paschi:

> o difesa di Dio, perché pur giaci?
> Del sangue nostro Caorsini e Guaschi
> s'apparecchian di bere: o buon principio,
> a che vil fine convien che tu caschi!
>
> (*Par.* XXVII. 55–60)[7]

The *Monarchia*, if it was composed towards the end of Dante's life, could not but share this spirit of indictment. It could not read otherwise than as a work of disillusionment and of repudiation.

So when was it written? In favour of an early dating is the evident dependence of the *Monarchia* on the *Convivio* in two respects. One is its wellnigh verbatim reproduction in Book II of *Convivio* IV's account of the providential course of Roman history—a product, as we have seen, of Dante's prophetic rereading of Virgil; and the other is its taking up of the *Convivio*'s distinction between present and future happiness, between the active happiness proper to this life and the contemplative happiness proper to the next life. The *Convivio*, it is true, advances only very tentatively along this road, and the question of papal-imperial relations does not crop up at all. Even so, there is no mistaking the kindred spirit of the *Convivio* and the *Monarchia* in their more 'humanist' or rationalist mood, the way in which the distinctions of the former look forward to those of the latter. It is as though, having lighted on a problem to be solved and on the means whereby to solve it (a parallel case, this time linguistic, is that of *Convivio* I and the *De vulgari eloquentia*), Dante pauses from his encyclopaedic work in the vernacular to settle there and then the burning issue. Between them, then, the shared Virgilianism and humanism of the *Convivio* and of the *Monarchia* point to an early dating for the *Monarchia*, to a date possibly as early as 1308–9, but probably, given its heightened sense of imperial promise, between 1310 and 1312.

But the manuscript tradition indicates a later dating: 1317 or 1318. Not that it dates the work specifically. It does not. What it does is to suggest as authentic the famous 'sicut in Paradiso Comedie iam dixi' clause of I. xii. 6 ('As I have already said in the *Paradiso* of my *Comedy*')—a reference to *Paradiso* V. 19–22 which, if authentic, would have the effect of putting back this part of the *Monarchia* to some time after the canto in question. True, there are difficulties here too, chief among them being that of reconciling the moral dualism of the *Monarchia*—its identification of *two* last ends for man—with the more unitary view of human experience proposed in the *Commedia*, with its taking up of all human activity in the final vision of God (a position much closer, for all Dante's differences with him on other scores, to that of Thomas Aquinas). But with the *Commedia* and the *Monarchia* representing as they do two very different kinds of undertaking—prophetic and ratiocinative respectively—the objection is not decisive; in contexts as diverse as this, basic structures, even basic ideological structures, may well differ. On balance, then, and given the priority of manuscript evidence over every other kind of evidence, be it stylistic, ideological, or circumstantial, it looks as though (with a twinge, perhaps, of misgiving) we must settle for a late dating.[8]

(ii)

Course of the argument

The doctrine of the *Monarchia* is simply stated. Dante begins by defining the end proper to mankind as a whole. This he does in terms of the intellect, or more precisely of the 'possible intellect' or total capacity in man for understanding. The intellect, as he explains in *Convivio* IV. xxi. 5, is open to receive every universal form. But since it is dependent for its operation on sense experience, which varies from individual to individual, it is only partially actualized in any one of them; hence the need for a multitude of men through whom the entire potentiality of the possible intellect may be realized (*Mon.* I. iii. 8). The end of mankind as a whole, therefore, is the collective actualization of the possible intellect, of man's total scope for understanding; and since he grows wise by living in peace (I. iv. 2),[9] it is the emperor's job to bring this about by making and promulgating good law.

With that, Dante passes to the arguments in favour of universal monarchy, some of which are based on the idea of unity and others on the idea of justice. The arguments based on unity run as follows. Everything *is* inasmuch as it approximates to oneness, to the ideal unity of its historical being. And society is no exception. The more it approximates to unity, the more stable it will be (v), the more orderly it will be (vi), and the more efficient it will be (xv). And this, Dante says, points to monarchy—indeed to a single universal monarchy—as the best form of political constitution, for only under a universal monarch will society be truly one. Indeed only under such a monarch will it work at all as a society, for only then will it resist its natural tendency to fragment. Universal monarchy, therefore, is of the essence. No universal monarchy, no peace, no progress, no real society. The argument from justice, an argument Dante had already developed in his letter (no. VI) to the 'iniquitous Florentines', runs as follows. Justice, Dante says, is a norm of activity permitting no deviation either way. It is absolute and unwavering. But inasmuch as it is represented by this or that individual it is mixed with its opposite, with greed and with the lust for power. Only in one whose jurisdiction is universal, and whose lust for power is thus satisfied, will it be pure, and the subject free to act in the light of conscience:

This understood, it then becomes clear how this liberty, or principle of all our liberty, is God's most precious gift to human nature, for, as I have already said in the *Paradiso* of my *Comedy*, by it we are made happy here as men, and happy as gods in the beyond. And if this is so, who will not agree that the human race is at its best when it is able to make the fullest use of this principle? But it is most free living under a monarch.

(I. xii. 6–7)[10]

Monarchy, then, is the guarantee of moral freedom. Here alone is there free passage from will to act, from seeing and understanding to doing.

Having thus argued his case on the basis of unity and of justice, all that remains for Dante to do is to clinch it historically by emphasizing (*a*) the fact of Christ's coming in the reign of the all-powerful Caesar Augustus—thus legitimizing that reign—and (*b*) the wretchedness of human society in moments of imperial interregnum. 'O humanity,' he concludes (I. xvi. 4–5), 'in how many storms must you be tossed, how many shipwrecks must you endure, so long as you turn yourself into a

THE *MONARCHIA*

many-headed beast lusting after a multiplicity of things! You are ailing both in mind and in heart. You minister neither to the higher part of the intellect by contemplating the unassailable truths of reason, nor to the lower by contemplating the face of experience, nor even to the heart by hearkening to the sweetness of divine counsel as breathed within you by the trumpet of the Holy Spirit: "Behold how good and pleasant it is for brethren to dwell together in unity" '—persuasion, characteristically, shading off into prophecy, into vatic denunciation.

In the second book Dante considers the question of Roman rule in particular, whether it was by right or by might that the Romans came to universal dominion. The question was widely debated. In Guelph, Angevin, and ecclesiastical circles the Augustinian view, that Roman rule was usurpation and that it was founded on violence, tended to prevail, and this, Dante says (II. i. 2), is a view he himself once took.[11] But now, having looked more deeply into the matter, he has come round to the kind of accommodation represented by, among others (though he does not name them), Tolomeo of Lucca—for whom the Romans were naturally fitted to rule—and Engelbert of Admont—for whom the whole of Rome's ascendancy was providential.[12] Everything about the empire, he says, bears the imprint of God's will, the sign of His favour resting on it. Take, for example, the nobility of the ancient Trojan line represented by Hector, Misenus, Aeneas, and Dardanus, and the selfless republicanism of Cincinnatus, Fabricius, Camillus, Brutus, Mucius, the Decii, and Cato (Cato especially); they are all perfect examples of the righteousness whereby the Romans came to world domination. Consider too the various miracles worked by God on Rome's behalf—the cackling geese and blinding hailstorm whereby she was saved from the ravages of the Gauls and of Hannibal (Chapter iv). And what of Rome's success on the battlefield, against, for example, the Rutuli or the Albans (Chapters viii–ix)? Does not this too testify to God's express intervention, to his guiding hand in the unfolding of Roman history? To think otherwise, Dante concludes (II. ix. 20), is foolish presumption. Indeed, it is to overlook the key conjunction in human history of the Roman and the Christian, the way in which Christ, living and dying as he did in the reign of Caesar Augustus, confirmed through Rome the universality of his atonement, his bearing of mankind's sins in their totality. The key passage, breath-taking in its conflation of the Roman and of the Judaeo-Christian in a single dynamic conception of human history, runs as follows:

> To be properly clear about this, it has to be understood that 'punishment' denotes, not simply the penalty inflicted on the wrong-doer, but that same penalty as inflicted by someone with the right to inflict it. Thus a penalty inflicted by anyone other than an official judge is no penalty at all. It should instead be deemed injury. That is why it was said to Moses, 'Who has appointed you judge over us?' Therefore, if Christ had not suffered under an official judge, that penalty would not have been punishment; and no judge could have been accounted official unless he had jurisdiction over the whole of mankind, since it was the whole of mankind that was to be punished in the flesh of Christ, who, as the Prophet says, 'bore our sorrows'.
>
> (II. xi. 4–5)

Rome's ultimate triumph, Dante is saying, and the ground of her legitimacy, lies in her facilitating the saving grace of Christ crucified. His appropriation of the classical for Christian purposes is complete.

Monarchia III is by any standard a remarkable book. Committed as it is to a humanist ethic pushed now to its limit, it displays on the methodological front a dialectical skill hitherto unprecedented even in the *Monarchia*. Gone is the subtlety and hyper-formalism of earlier passages in the work (of, for example, I. xi. 9 or xiv. 3). Instead, there is a fresh directness of manner as Dante proceeds first to combat his opponents' position and then to state his own. The prelude is distinctly embattled: 'I will now enter the arena,' Dante declares, 'and in the strength of him who liberated us from the powers of darkness by his very blood, and before the gaze of the whole world, fling from it the impious and the liar' (III. i. 3). The basic question is this: is the authority of the emperor dependent on God or on the pope? Does it flow directly from Him or indirectly by way of Peter and of his successors to the keys? (i. 5). The question has, as we shall see, a long and complicated history, but it is enough for Dante's purposes in the *Monarchia* to identify the current opposition. First, he says, there is the pope—led astray by his zeal for the keys (III. iii. 7)—together with other over-zealous pastors of the Church. Then there are the massed ranks of (largely Guelph) princes, jurists, and propagandists opposing the emperor's claims to parity and independence *in temporalibus*—sons of the devil, Dante says, one and all. And then there are the decretalists, commentators, not on the most ancient and authoritative documents of the Church (the Scriptures), but on the successive and mostly *ad hoc* pronouncements of successive popes bent on vindicating papal power in worldly affairs. They too are beneath consideration and should be 'chased from the arena' (III. iii. 11). His argument is with the first group, with the zealous but misguided prelates (iii. 18).

Chapters iv–xi consider the various kinds of argument advanced by the papal theorists in favour of the dependence of secular on spiritual authority. There are three types: scriptural-exegetical, historical, and (for want of a better term) relational—to do with how one kind of authority relates to another. The scriptural-exegetical type has to do with passages and motifs from the Bible and with what, if anything, they may be said to yield allegorically. Examples adduced by Dante include the Genesis account of the two luminaries (the sun to preside over the day and the moon over the night), the offices of priest and king associated with Levi and Judah (Genesis 29: 34–5), the election and deposition of Saul by Samuel in 1 Samuel 10 and 15, Matthew's account of the Magi's gifts of gold and frankincense (2: 11) and of the 'whatsoever you bind' commission of Christ to Peter (16: 19), and the two swords episode in Luke 22. These and others like them were the classical *loci* of theocratic scriptural exegesis, each in its way confirming, or so it was maintained, the subservience of the secular to the spiritual, of the princely to the priestly. To them Dante brings a nice sense of logical and interpretative correctness. His case, he says, will rest (*a*) on the exposure of material and formal error, and (*b*) on a denial of allegorical intent in the first place. Thus the luminary argument, equating the sun with papal authority and the moon with imperial authority, is faulty both materially—in that the sun and the moon were created prior to man, and could not, therefore, be intended as symbols of human authority—and formally—in that it involves a faulty association of light and authority. The same applies to the Levi and Judah argument, to the notion that the empire is related to the Church as Judah is related to Levi. It is faulty both materially—in that there is no basis for linking Jacob's two sons with the two forms of government—and formally—in that the predicate differs in meaning from major premiss to the conclusion (seniority and authority). As for the arguments from Matthew and Luke, it is enough in the case of the latter to deny allegorical intention at all, and in the case of the former to point up the mistake of identifying Peter's authority and Christ's. The two are not the same, for Peter stands to Christ, not as equivalent or plenipotentiary, but as appointee to a particular order of responsibility. 'Being a man's vicar', Dante says (III. vii. 7), 'does not make the vicar equivalent to the man himself.' On the contrary, it confirms the difference between them. And so it goes on, each argument in turn being carefully dismantled in order to expose its misconception and misuse.

Dante turns next to the type of argument based on historical precedent. To the fore is the Donation of Constantine, though he also considers Charlemagne's coronation by Pope Adrian (in fact Leo III) in the year 800. The Donation, a document purporting to bestow on the Church dominion over the western empire, occupied a key place in hierocratic thinking. It was not until the fifteenth century and Lorenzo Valla that it was finally denounced as a forgery, but long before this doubts had been expressed as to its legal validity. This was Dante's position. The Donation, he believed, well-intentioned as it was, was misconceived, for it set out to alienate the ground of its own authority:[13]

> Again, all jurisdiction is prior to its judge, since the judge is appointed to the jurisdiction, not vice versa. But the Empire is that jurisdiction which comprehends within it every temporal jurisdiction. Therefore it is prior to the judge, that is, to the emperor, he being appointed to it rather than it to him. Clearly then, the emperor as Emperor cannot alter it, since it is by virtue of his jurisdiction that he is what he is.
>
> (III. X. 10)

But doubts on the legal front coalesce with urgency on the reform front, with a sense of the Church's having forsaken the Gospel notion of poverty (Matt. 10: 9). 'Searching diligently in the scriptures,' Dante says (III. xiii. 4–5), 'I have looked in vain for any command laid upon priests either of the old or of the new dispensation to oversee temporal affairs. On the contrary, I find that the priests of the old dispensation were specifically excluded from temporal affairs by God's commandment to Moses, and those of the new dispensation by Christ's command to his disciples.' Christ's kingdom is not of this world. It is not a kingdom like others, mundane, wealthy, and domineering. It is a kingdom of the spirit, indifferent to the substance and trappings of temporal government. Dante's final position, then,—a position responsive to the promptings of a reformist or 'spiritualist' conscience—is richly nuanced: the empire was no more in a position to make the Donation than the Church was to receive it. Each in doing so has compromised its *raison d'être*, its proper nature and finality.

Finally, there is the 'relational' argument (Chapter xi). Everything, the hierocrats maintain, is subject to whatever is best in its own species, to the *signum simplicissimum* normative for all. In the case of mankind this means subordination to the pope, who, 'judging all men and judged by none' (1 Cor. 2: 15), is the best man. This, Dante says, is easily dealt with. It is a question of categories and of relationships. Categorically, the terms 'pope' and 'emperor' designate, not a substance, but an accident. They denote offices and responsibilities—what a man *does* rather than what he *is*. 'Pope' denotes a relationship of fatherhood (*paternitas*), while 'emperor' denotes a relationship of law and coercion (*dominatio*). The two are not the same. Indeed, they are mutually irreducible. While, then, pope and emperor *as men* are reducible to the best man as defined by Aristotle in the *Ethics*, they are not reducible to a common principle *qua* pope and emperor—except, of course, to God, in whom all things come together.

With the hierocratic case thus exposed for what it is—an amalgam of misconception and logical solecism—Dante sets about putting his own case, about defining the relationship as *he* sees it between papal and princely power. Papal and princely authority, he believes, exist in parallel. They are not dependent one on the other. They are each directly dependent on God. Thus the argument is staunchly dualist, fervently impatient of any kind of mutual reductionism. Let us follow it step by step.

Man, Dante says, tracing a pattern of thought with which we are now familiar, is a twofold creature of flesh and spirit. He stands on the horizon between the material and the spiritual world, participating in both. This means that he has, not one, but two ends ('duo ultima'), one in respect of his material and corruptible nature, and one in respect of his immaterial and incorruptible nature, the 'material' end being realizable in time and the 'immaterial' end in eternity (III. xv. 5–6). But now Dante shifts his ground; for what in truth he wishes to define and to vindicate is the happiness here and now, not of the body (for happiness in this sense is strictly limited), but of one aspect of man's *spiritual* activity, the aspect represented by the 'natural' moral and intellectual virtues of justice, fortitude, prudence, and so on over and against the 'infused' theological virtues of faith, hope, and love. The former, he says, supervised in their practical working out by the emperor and the philosopher, are productive of perfect happiness here and now—of the happiness of the 'earthly paradise'—while the latter, supervised by the pope, are productive of perfect happiness hereafter— the happiness of the 'heavenly paradise'. The key passage, III. xv. 8–10, runs as follows:

These two sorts of happiness are attained by diverse means, just as one reaches different conclusions by different means. We attain to the first by means of philosophical teaching in so far as we follow this by exercising the moral and intellectual virtues, while we arrive at the second by means of spiritual teaching (which transcends human reason) in so far as we follow this by exercising the theological virtues of faith, hope, and love. These conclusions, and the means towards them, have on the one hand been revealed to us by human reason, which through the philosophers has enlightened us on every issue, and on the other hand by the Holy Spirit, who has made known the supernatural truth necessary for our salvation by means of the prophets and of the sacred writers, and through the Son of God, who is co-eternal with the Spirit, Jesus Christ, and through Christ's disciples. Even so, human cupidity would fling aside such aids if men, like horses wandering freely to satisfy their bestiality, were not held to the right path 'by the bit and the rein'. This explains why two guides have been appointed for man to lead him to his twofold goal: there is the supreme Pontiff, who is to lead mankind to eternal life in accordance with revelation; and there is the emperor, who, in accordance with philosophical teaching, is to lead mankind to temporal happiness.

THE *MONARCHIA*

With this, Dante has achieved his purpose in the *Monarchia*. Man, he has shown, has two natures, two ends, two means, and two lords, the emperor (assisted on the moral and intellectual side by the philosopher, an association surviving from *Convivio* IV. vi. 17–18) and the pope. Each has his own jurisdiction, these areas of jurisdiction existing in parallel. True, a caveat is added at III. xv. 16–18. There is, Dante suggests, a hierarchy, if not of domination then of dignity, linking pope and emperor. Caesar, he says, related as he is to Peter in the way a first-born son is related to his father, receives from him the light of grace and benediction 'ut . . . virtuosius orbem terre irradiet' ('that he, Caesar, may the more powerfully enlighten the world'). But, as we have already discovered from the 'relational' argument of Chapter xi, dignity—in the sense of fatherly dignity—is not domination, and so the principal thesis stands undiminished. *Imperium* and *sacerdotium* both derive directly from God and exercise their functions independently of each other.

(iii)

Papal hierocracy, moral finalism, and the Monarchia *as a work of Christian political philosophy*

The *Monarchia*, we began by saying in this chapter, is open to interpretation in two different but related ways: (*a*) as an occasional piece of publicistic literature, and (*b*) as, in and through the publicistic, an essay in Christian humanism. The phrase 'in and through' is important, for though it is mainly along the lines of (*b*) that we shall develop our argument here—in terms of the *Monarchia* as further testimony to the strain of Peripatetic rationalism in Dante—all this, in the text, is at the service of the political. Here, then, with the practical problem of popes and emperors, is where we shall begin.[14]

The problem went back a long way, as far as the beginning of the fourth century, when the Emperor Constantine, in the Edict of Milan (313), recognized Christianity as the official religion of the empire. At long last the days of oppression and persecution were over, and Christians were confronted now by the problem, not of escape and of dissimulation, but of establishment, of coming to terms with the prevailing structures. And this meant sorting out the question of authority; for while it was clear that the emperor was supreme *in temporalibus*, in matters pertaining to man's well-being in time and space, it was equally clear that this was an order of authority transcended by the pope's concern, not with time, but with eternity, with man's immortal end. That in the long run was what mattered. Everything else was provisional, merely along-the-way. So how should the temporal authority relate to the spiritual? Gelasius I, towards the end of the fifth century, hit upon a nicely even-handed solution. Pope and emperor, he said, should coexist as equal partners in human government. The pope's perspective was longer, but as far as this life was concerned a proper apportioning of roles and responsibilities was both desirable and possible. The pope would look after the moral and spiritual aspects of man's life, thus ensuring his immortal welfare, while the emperor would look after the constitutional, legal, and military aspects, thus ensuring his mortal welfare.[15] And this, in times of uncertainty and indeed of catastrophe for the empire, worked well, the bishops, and the bishop of Rome in particular, taking an if anything more than merely spiritual responsibility for the well-being of their subjects. But with the increasing wealth and centralization of the Church as the Middle Ages wore on, and with the growth of nationalism in Europe, especially in France and England but also, in a less coherent way, in the German world, tensions began to develop, secular princes and princelings being reluctant to knuckle under to Rome and to the papacy. The first great trial of strength was the so-called Investiture Contest of the 1070s, a complex affair turning on the right of episcopal conferment. Henry IV, having taken it on himself to invest the archbishop of Milan, found himself in deep water. A victim to the catch-all clause of Matthew 16: 19—the 'whatsoever you bind' clause—he found himself deprived of all dominion in Germany and Italy.[16] In the event, he thought better of it, submitting himself humbly, indeed humiliatingly, for forgiveness at Canossa, and, for a time, peace reigned, Gregory VII basking in the completeness of his triumph. But what with the now inveterate German mistrust of Roman priestliness and, more concretely, the rise of Frederick I (Barbarossa) and Frederick II of Hohenstaufen ('stupor mundi') in the middle part of the next century and in

THE *MONARCHIA*   161

the early part of the thirteenth century, the quarrel over the two powers was set fair to continue. And it was in Dante's time, under Boniface VIII, that, again over a practicality, it came to a head. Philip IV of France, at war with Edward I of England, wanted to tax the French clergy to raise funds. But the pope, jealous of his transalpine revenue, would have none of it, and, after a series of dire pronouncements in the months preceding, he promulgated in November 1302 his bull *Unam sanctam*, a neat if predictable florilegium of traditional hierocratic motifs. The faith, Boniface said, was one, the Church was one, baptism was one, and authority was one. Outside the Church, which meant outside the authority of Peter and of Peter's successors, there was no salvation. Let all princes and princelings, therefore, bend the knee. Let them as a matter of eternal consequence submit humbly to the oversight of God's highest appointee on earth:

In the Church and in her power are, as we learn from the words of the Gospel, two swords, the spiritual and the temporal . . . both are in the power of the Church, both the spiritual and the material sword; but the latter is to be exercised on behalf of the Church, the former *by* the Church, the former by the hand of the priest, the latter by the hands of kings and soldiers but at the bidding and by the forbearance of the priest . . . the temporal power ought to be subject to the spiritual power . . . for as truth itself testifies, it belongs to the spiritual power to institute the earthly power, and if it be not good to judge it. Whosoever, therefore, resists this power ordained of God, resists the ordinance of God . . . Furthermore, we declare, state, define, and pronounce that for every human creature to be subject to the Roman pope is altogether necessary for salvation.[17]

Now the pope, it has to be said (lest we rush into easy judgements about ecclesiastical totalitarianism), had a point here. He had a point in that papal theocracy *as an idea*, as a way of thinking about the Christian community as a whole, was a good one. Indeed, rooted as it was in the deepest movements of the medieval conscience, it was wellnigh incontrovertible. Theologically, it reflected the unity and simplicity of God. God is one and undivided, and that too is how His Church should be, single and undivided in truth and authority. Philosophically, it answered to the Christian Neoplatonic sense of the *ens simplicissimum*, of the transcendent form manifest in, but not coincident with, the empirical phenomenon here below, with the Church militant on earth. Constitutionally, it chimed well with Roman legal corporationism, with the notion of the Church as a *persona ficta* variously endowed with rights, privileges, and immunities. Historically, it bore about it the accumulated wisdom of the ages, the sanction of one sacred pronouncement after another. Morally, it defined with perfect precision the arena of human activity generally, the long-term as well as the immediate perspective of man's doing and deciding as a rational creature. Psychologically, it offered guidance and security against the insidious forces of loss, perplexity, and alienation; while aesthetically it manifested a perfect coming together of the parts in the whole, a perfect unity in diversity. What, then, was Dante's problem? How could he oppose an idea of such impeccable pedigree and deep necessity?

He opposed it for two reasons. In the first place, there was his growing sense—ultimately Franciscan in origin—of the Church's betrayal of its evangelical roots, of the way in which, by laying claim to power and wealth, it had sold short the Gospel injunction to poverty, the Christian sense of an inward, and inwardly reforming, kingdom of the spirit. And this, as time goes on, becomes the dominant consideration in his mind.[18] But alongside it, and, as far as the *Monarchia* is concerned, as decisive, is his sense—this time Aristotelian in origin—of the blandness of contemporary theology, of its easy taking up of the mortal in the immortal, of time in eternity, of present in future possibility; for at the heart of papal theocracy as contemplated by Dante in the *Monarchia* was a deeply ingrained sense of transcendent moral finalism. Everything man is and everything he does is directed towards the one transcendent end. Everything he imagines, contemplates, plans, and seeks to realize stands to be judged in the light of the last, transcendent, immortal goal (mediated, needless to say, by priests). Take, for example, the *De ecclesiastica potestate* of Giles of Rome (or Egidio Colonna as he is also known).[19] This, fundamentally, is an essay in moral and aesthetic unity. Creation, Giles says, bearing as it does the imprint of God's grand design, is perfectly ordered. To suppose that parts of the whole can be hived off to form a separate order and jurisdiction is entirely untoward ('omnino inconsequens'):

THE *MONARCHIA*

For which reason, if the whole of the universe, of which God has overall care, is so well ordered that inferior bodies are subject to superior bodies, that the bodily generally is subject to the spiritual, and indeed that things spiritual are subject to the supreme spirit—that is, to God—to declare that God's faithful people, and indeed the Church itself, chosen by God for His own, spotless and without blemish, is not well ordered, that it is not bound harmoniously one part with another, and that the order of the universe as a whole, which is an order of the greatest beauty (indeed, as Augustine has it in the *Enchiridion*, of a most awesome beauty), does not shine forth in the Church, is altogether untoward.

(*De eccles. pot.* I. v)

The spiritual and the temporal are not separable, he goes on. Rather, the spiritual inheres in the temporal as its inmost principle and *raison d'être*. The temporal *is* the spiritual, the two belonging together within the economy of the whole:

In all these ways the temporal might be deemed spiritual; indeed, it was in the light of all this that things temporal were first said to fall within the jurisdiction of the spiritual power.

(ibid. III. v)

—an emphasis making straight away for theocracy, for oversight of human activity generally by those concerned with its highest end.

James of Viterbo, whose *De regimine christiano*, like the *De ecclesiastica potestate* of Giles of Rome, dates from about 1302, is another case in point. James, unlike Giles, starts out Peripatetically, insisting, not on the institution of the secular by the spiritual (cf. *De eccles. pot.* II. v: 'for kingly power is instituted for and by the Church'), but on the notion of society as a product of man's natural gregariousness: 'The institution of these communities or societies comes about, as Aristotle shows in the first book of his *Politics*, as a matter of human instinct', he says in his first chapter. Conscious, however, of the inbuilt centrifugalism of this view, of its tendency to spawn a variety of independently-minded secular authorities, he quickly hurries over to the Augustinian pole of his inspiration. There can, he says (II. vii), be no true virtue where there is no true faith, and no true or legitimate authority where there is no oversight by the Church. The Church, therefore, even given the naturalistic emphases with which James starts out, retains her accustomed privileged position, having the right—all in the name of man's last end and immortal happiness—to sanction, to amend, and even to do away with provisions of the secular authority.

But by far the most interesting case of moral finalism at work in the political sphere is that of Thomas Aquinas (d. 1274).[20] Thomas, by instinct, is a secularist. For him, the secular, left to itself, is capable of taking care of itself. It does not need the intervention of priests. All it needs is expert lawmakers and judges, men of professional skill, of good will, and of sound conscience (*ST* ia iiae 90 ff.). Problems arise, however, when Thomas turns to the question of spiritual and secular power proper, for here the finalism of the Christian conscience has the last word. There are two passages to consider, one from the early *Commentary on the Sentences of Peter Lombard* (ii. 44. 3. 4) and the other in the incomplete treatise—taken up and developed along hierocratic lines by Tolomeo of Lucca—*De regimine principum*. The first passage is a curious, and (reading between the lines) somewhat reluctant, nod in the direction of papalism, the 'nisi forte' ('unless perhaps') clause towards the middle qualifying the dualism of what precedes in a distinctly lacklustre, formulaic fashion:

Both the spiritual and the temporal power derive from the divine power; consequently the temporal power is subject to the spiritual power only to the extent that this is so ordered by God—namely in those matters which affect the salvation of the soul. And in these matters the spiritual power is to be obeyed before the temporal. In those matters, however, which concern the civil welfare, the temporal power should be obeyed rather than the spiritual, according to what we are told in St Matthew (22: 21): 'Render to Caesar the things that are Caesar's'. Unless, perhaps, the spiritual and the temporal power are identified in one person as in the pope, whose power is supreme in matters both temporal and spiritual, through the dispensation of Him who is both priest and king; a priest for ever after the order of Melchizedek, the king of kings and lord of lords, whose power shall not fail and whose dominion shall not pass away to all eternity. Amen.

But the second passage, from i. xiv of the *De regimine*, is more definite. Addressed to the king of Cyprus and inclined by way of checks and balances to take a clerical view, it focuses on the question of ends. Temporal affairs, it is true, remain distinct from spiritual affairs, and have their own highest authority, the king, to whom all others in the secular sphere are subservient. But ends are ends, and man's final end, which lies in the enjoyment of God not here but hereafter, is overseen not by princes but by priests, and above all by the pope, 'to whom all are subject as to the Lord Jesus Christ himself':

Now the man who lives virtuously is destined to a higher end, which consists, as we have already said, in the enjoyment of God; and the final object of human association can be no different from that of the individual man. Thus the final aim of social life will be, not merely to live in virtue, but rather through virtuous life to attain to the enjoyment of God. If, indeed, it were possible to attain this end by human virtue, it would, in consequence, be the duty of kings to guide men to this end. We believe, however, that it is the supreme power in temporal affairs which is the business of a king. Now government is of a higher order according to the importance of the end it serves. For it is always the one who has the final ordering of affairs who directs those who carry out what pertains to the achievement of this final aim; just as the sailor who must navigate the ship advises the shipwright as to the type of ship which will suit his purpose, and the citizen who is to bear arms tells the smith what weapons to forge. But the enjoyment of God is an aim which cannot be attained by human virtue alone, but only through divine grace, as the Apostle tells us (Romans 6: 23): 'The grace of God is eternal life.' Only a divine rule, then, and not human government, can lead us to this end. Such government belongs only to that King who is both man and also God, that is to Jesus Christ, our Lord, who, making men to be sons of God, has led them to the glory of heaven.

This, then, is the government entrusted to him, a dominion which shall never pass away, and in virtue of which he is called in the Holy Scriptures not only a priest but a king, as Jeremiah says (23: 5): 'A king shall reign and shall be wise.' It was from him that the royal priesthood derives; and, what is more, all the faithful of Christ, being members of him, become thus priests and kings. The ministry of this kingdom is entrusted not to the rulers of the earth but to priests, so that temporal affairs may remain distinct from spiritual; and in particular it is delegated to the High Priest, the successor of Peter and Vicar of Christ, the Roman Pontiff; to whom all kings in Christendom should be subject, as to the Lord Jesus Christ himself.

Again, there is something formulaic about the passage—certainly, at any rate, about the last part with its grandiose

'Summo Sacerdoti, successori Petri, Christi Vicario, Romano Pontifici . . .'—something of the standard clause. But the standard clause is in this case all of a piece with the continuities of Thomist moral philosophy and epistemology. Man, Aquinas insists, is an intellectual creature in the flesh and destined, as his proper operation and highest vocation, to *know*. The range of his understanding, or more precisely of his understanding based on observation and reasonable inference, is wide, as is his scope for moral self-determination. But Thomas's reading of the philosophers, and in particular of *the* Philosopher (Aristotle), is located within the revelatory context of Christianity, which means the lifting and completing of human knowing and doing through the grace of God. In the end, therefore, distinction gives way to succession, legal secularism to moral finalism, and political dualism to theocracy.

To this, Dante, as we have seen, brings a characteristic separating out of the 'human' and the 'more-than-human', of the here and now and the hereafter. But—and this now is what matters—where before, in the *Convivio*, he had been content to proceed in terms of the moral, in terms of what man *does*, here in the *Monarchia* he proceeds in terms of the intellectual, in terms of what man *knows*—this, in the circumstances, making all the difference between mild eccentricity and out-and-out heterodoxy. There are two passages to consider, I. iii. 4–10 and III. xv. 7–15. Any discussion of imperial power, Dante begins in I. iii, must begin with its overall purpose, which is to assist in bringing about man's proper end as man, his specifically intellectual well-being. More precisely, it is there to secure the peace and quiet necessary to bring about man's *collective* intellectual well-being, his shared capacity for understanding:

> And because this potentiality cannot wholly and at once be translated into action by one man, or by any one of the particular communities identified above, there must needs be a multiplicity in the human race whereby the entire potentiality can be actualized . . .
>
> (I. iii. 8)

Now Dante, having mentioned the actualization of man's total capacity for understanding, found himself in hot water, this, to some, looking suspiciously like Averroism (outlawed in theological circles for its tendency by way of monopsychism to deny the possibility of personal immortality). Guido Vernani especially, whose *De reprobatione Monarchie Dantis* appeared in 1327,[21] became very agitated over precisely this point. Dante, he said, having spoken of collective understanding, and having, moreover, cited Averroes by name (I. iii. 9), had committed the worst possible of all errors:

> The third error is the worst, for he then says, in the same chapter and in the one following, that the possible intellect cannot be realized, that is perfected, other than through the whole human race, just as the potential of prime matter cannot wholly be reduced to act and perfection other than through a multitude of natural phenomena. And to this end he adduces the authority of Averroes in the third book of his *Commentary on the De anima*. And this, he says, is the end and perfection, not of one single man, but of the whole human race taken together. Speaking in this way, it follows clearly that in all men there is but one single intellect—to say and suppose which is the worst possible error, whose author and founder was the very Averroes whom he adduces.
>
> (*De rep. Mon.* I; ed. Matteini, 97)

In the event Vernani was wrong, for in the I. iii. 8–9 passage to which he refers Dante is talking, not about the nature of the human intellect in itself, but about understanding *in the aggregate*, as a shared endeavour. Never, in fact, for all his enthusiasm for Averroes, did he consent to the notion of monopsychism. Other things—including the idea of a speculative happiness here and now—yes, but not that. But in respect of the III. xv passage, the one about two natures and two ends, Vernani was closer to the mark. Dante, it will be remembered, is anxious in this final chapter of the *Monarchia* to demonstrate the independence of imperial authority with respect to papal authority, to which end he draws a distinction between man's mortal nature and his immortal nature, between his mortal happiness and his immortal happiness, between moral and intellectual virtue and theological virtue, and, as his point of arrival, between imperial and papal jurisdiction. The key lines are these:

> These two sorts of happiness are attained by diverse means, just as one reaches different conclusions by different means. We attain to the first by means of philosophical teaching in so far as we follow this by

exercising the moral and intellectual virtues, while we arrive at the second by means of spiritual teaching (which transcends human reason) in so far as we follow this by exercising the theological virtues of faith, hope, and love.

(III. xv. 8)

to which Vernani replied as follows:

He says, moreover, that man is ordered to these two ends by God. To this I reply that man is not directed by God to temporal happiness as to his last end, since never has happiness of this kind been enough to still and to satisfy his desire; on the contrary, even philosophically speaking, the exercise of these virtues is ordered to a contemplative happiness, such that, having allayed all other passions, he may the more freely and peacefully contemplate, in wisdom, the things of eternity ... man, therefore, is ordered to eternal happiness as to his last end, to the pursuit of which all goods, natural, moral, and gratuitous, should be referred and directed.

(*De rep. Mon.* III; ed. Matteini, 117)

Now Vernani was no genius. Indeed, to go by (the admittedly combative) Bruno Nardi, he was a cantankerous old pedant—a 'malevolo teologo'.[22] Here, though (as Gilson noted) he has a point, for in the last chapter of the *Monarchia* we witness the extreme limit of Dante's rationalism. Cornered by the theologians, but prompted even so by his own deepest instincts, he resorts at every turn to distinction and opposition. On the one hand the mortal, on the other the immortal. On the one hand present happiness, on the other future happiness. On the one hand the immanent virtues of prudence, justice, fortitude, and courage; on the other, the infused virtues of faith, hope, and love. On the one hand the natural light of reason; on the other, the gracious light of revelation. Nowhere is there any coming together of these two poles of Dante's awareness, any notion of human activity as justified *in its totality* in and through the Christian. Nowhere is there any sense—so fundamental to the *Commedia*—of the old as transcended by the new, as subject in Christ to inward transformation. Instead, and as far, at any rate, as this life is concerned, all is Aristotelian non-self-transcendence, classicism, we might say, undisturbed by any deeper and more far-reaching sense of human possibility.

That, though, is not the whole picture, for if in one sense the

THE *MONARCHIA* 169

Monarchia represents (in Gentile's words) 'an act of rebellion against scholastic transcendence',[23] in another it remains firmly rooted in scholastic and medieval habits of mind. One of these, the book's profound sense of providentialism, we have already touched on in connection with the *Convivio* and in our summary of *Monarchia* II. Roman history, Dante thinks, subsists alongside Jewish history as an equally essential part of God's purposes for man. From the beginning it was divinely directed towards the climactic moment represented by Christ's coming in the reign of Caesar Augustus, there to make plain once and for all His ideal pattern of human government. But alongside the providential is the Augustinian, and this is what really interests us; for Augustinianism, while prepared to acknowledge the place of Rome in the divine plan, sits most uncomfortably with Aristotelianism. Aristotle, standing back to contemplate man and society, took a genial view. It was all, he thought, a matter of natural evolution, of men's coming together in groups for the sake of the common good. First come neighbourhoods and small communities, then villages and cities, and finally whole states and kingdoms, the mechanisms of government, be they monarchical, oligarchical, or democratic, being there simply to provide the necessary checks and balances, the necessary dose of pure positive law. But for Augustine, whose political circumstances differed from those of Aristotle, political authority was not so much normative as coercive. It was there, not so much to foster and guarantee the smooth working of the social machine, as to control what Augustine saw as man's congenital unruliness. In short, it was a remedy for sin, a response to the fact of man's vitiation at the time of Eden. Whereas, then, the Aristotelian-Thomist citizen, being a man of moral substance, needs a minimum of constraint, and should not be fussed by the law, the Augustinian citizen needs little else. Deeply perverse, he has constantly to be kept in order.

Now in Dante, for all his confidence in the notion of happiness here and now as a consequence of moral and intellectual virtue, there remains much of the Augustinian, much of the despairing in respect of man's ever making good without the discipline of *imperium*. We catch a glimpse of it in, for instance, the following passage from the *Convivio* (IV. ix. 8 and 10), where thoughts on the necessity of imperial constraint are linked explicitly with the name of Augustine and with the great bishop's inkling of man's depravity:

> Onde dice Augustino: 'se questa—cioè equitade—li uomini la conoscessero, e conosciuta servassero, la ragione scritta non sarebbe mestiere' . . . Sì che quasi dire si può de lo Imperadore, volendo lo suo officio figurare con una imagine, che elli sia lo cavalcatore de la umana volontade. Lo quale cavallo come vada sanza lo cavalcatore per lo campo assai è manifesto, e spezialmente ne la misera Italia che sanza mezzo alcuno a la sua governazione è rimasa!

The drift of Dante's thought is clear: if men knew the meaning of equity—i.e. of the right in respect of voluntary activity ('volontarie operazioni')—and acted accordingly, then there would be no need of law. But they don't and there is, and that is why human nature has, like a beast, to be ridden and reined in. But the most spectacular instance of political Augustinianism in the *Convivio*, 'spectacular' in the sense of spectacularly contrastive, comes a little earlier, in IV. iv. 1–4. Richly various in inspiration, the passage is worth quoting in full:

> Lo fondamento radicale de la imperiale maiestade, secondo lo vero, è la necessità de la umana civilitade, che a uno fine è ordinata, cioè a vita felice; a la quale nullo per sé è sufficiente a venire sanza l'aiutorio d'alcuno, con ciò sia cosa che l'uomo abbisogna di molte cose, a le quali uno solo satisfare non può. E però dice lo Filosofo che l'uomo naturalmente è compagnevole animale. E sì come un uomo a sua sufficienza richiede compagnia dimestica di famiglia, così una casa a sua sufficienza richiede una vicinanza; altrimenti molti difetti sosterrebbe che sarebbero impedimento di felicitade. E però che una vicinanza non può sé in tutto satisfare, conviene a satisfacimento di quella essere la cittade. Ancora: la cittade richiede a le sue arti e a le sue difensioni vicenda avere e fratellanza con le circavicine cittadi; e però fu fatto lo regno. Onde, con ciò sia cosa che l'animo umano in terminata possessione di terra non si queti, ma sempre desideri gloria d'acquistare, sì come per esperienza vedemo, discordie e guerre conviene surgere intra regno e regno, le quali sono tribulazioni de le cittadi, e per le cittadi de le vicinanze, e per le vicinanze de le case, e per le case de l'uomo; e così s'impedisce la felicitade. Il perché, a queste guerre e le loro cagioni tôrre via, conviene di necessitade tutta la terra, e quanto a l'umana generazione a possedere è dato, essere Monarchia, cioè uno solo principato, e uno prencipe avere; lo quale, tutto possedendo e più desiderare non possendo, li regi tegna contenti ne li termini de li regni,

sì che pace intra loro sia, ne la quale si posino le cittadi, e in questa posa le vicinanze s'amino, in questo amore le case prendano ogni loro bisogno, lo qual preso, l'uomo viva felicemente; che è quello per che esso è nato.

Imperial authority, Dante starts out by saying, is a necessity of civilized existence. Like every other form of political authority it is rooted in social evolution, in the ever-increasing need for collective self-help. First comes the family, then the neighbourhood ('vicinanza'), then the city, and then the kingdom, each designed to remedy the inadequacy of the one preceding—all very Aristotelian ('E però dice lo Filosofo . . .') and naturalistic. But then, at the level of *imperium*, there is a sudden change of direction, evolutionism giving way to the notion of authority as corrective, as a check to men's perennial thirst for glory and self-aggrandizement. Look round for yourself, Dante says ('sì come per esperienza vedemo'), and there it is: greed, strife, discord, and misery; hence the need for an empire, for a force in human affairs designed to keep the peace and to constrain man to his proper happiness. And this fundamentally pessimistic view of human nature, this sense of the decisiveness of the Fall, impinges on the *Monarchia* to check its otherwise all-dominating sense of man's natural moral and intellectual viability. A first indication comes in III. iv. 14 in the course of Dante's refutation of the two luminaries argument adduced by the hierocrats—the notion that the sun, the brighter of the two, equals the papacy, and the moon, the lesser of the two, the empire. Not so, he says, since man, for whom these two powers represent a remedy for sin—'remedia contra infirmitatem peccati'—had not yet been made. 'He would indeed be a stupid doctor', he goes on, 'who prepared a plaster for one not as yet born', imperial authority thus emerging, not as normative, but as curative, as means of making good. Final confirmation, however, comes in the wake of the III. xv passage cited above, the passage about the twofold means and ends of human happiness. Unconstrained, Dante says, man would fling aside all aids to his present and future well-being in favour of greed and bestiality. Overrun by the 'waves of alluring cupidity' ('fluctibus blande cupiditatis'), he would rush headlong to his self-inflicted annihilation:

... human cupidity would fling aside such aids if men, like horses wandering freely to satisfy their bestiality, were not held to the right path 'by the bit and the rein'. This explains why two guides have been appointed for man to lead him to his twofold goal: there is the supreme pontiff, who is to lead mankind to eternal life in accordance with revelation; and there is the Emperor, who, in accordance with philosophical teaching, is to lead mankind to temporal happiness. But since none, or very few at least, and then only with the greatest difficulty, would reach this harbour unless the waves of alluring cupidity be assuaged and mankind freed from them so as to rest in the tranquillity of peace, this is the object to which, above all others, he who has charge of the world as a whole must devote his energies who is called the Roman Prince, that on the threshing floor of this life man might live in freedom and in peace.

(III. xv. 9–11)

What now, then, of Dante's humanism, of his sense of man's reaching out in his own strength for happiness in this life? The answer is that it is still there, still alive as one of the poles of his inspiration. Caught up, however, between the splendour of the Aristotelian way and the tragic persuasiveness of the Augustinian way, it is at once qualified by the alternative view, by the more recognizably Christian sense of man's standing in need *even now* of saving grace.[24]

Considerations of this sort, however, though they qualify the 'subversiveness' of the book, its value as 'an act of rebellion against scholastic transcendence', are marginal to its main thrust, to the principles on which Book III in particular rests; for central to Book III is its sense of imperial authority, not as a remedy against sin, but as a guarantee of man's making good 'ungraced', by means of natural moral and intellectual virtue. Sin impinges, but does not waylay the argument. Here, though, lies perhaps the most intriguing aspect of the *Monarchia*; for this separating out of nature and grace emerges at last to trouble Dante's own conscience. The key passage is the famous and controversial caveat of III. xv. 17–18, the last lines of the book. It runs as follows:

The truth on this issue is not to be so narrowly interpreted as to exclude the Roman Prince altogether from subordination to the Roman Pontiff, since in a certain fashion our temporal happiness is subordinate to our eternal happiness. Let Caesar, therefore, observe that reverence towards Peter which a first-born son owes to his father; so that, enlightened by paternal grace, he may the more powerfully enlighten the world, at the head of which he has been placed by the One who alone is ruler of all things spiritual and temporal.

The passage has been variously interpreted. It has even been seen to represent a tardy nod in the direction of *potestas indirecta*, a last-minute concession to the papal claim to casual intervention in the temporal order.[25] The notion is impossible, intervention of this kind being everywhere repugnant to Dante, starkly counter to his every instinct. If there is a concession—and this brings us back for the last time to the principal theme of this chapter—it is at the moral, or rather existential, level; for whatever the dialectical value of Dante's leading distinctions—of the importance of distinguishing between the reasonable and the revealed as a principle of debate (and no one was in doubt about that)—existentially, and on any specifically Christian reading of the matter, the individual stands *at once*—not consecutively but *contemporaneously*—in time and eternity. He lives, moves, and has his being in both dimensions. To suppose otherwise is seriously to prejudice our understanding both of man and of God: of man in respect of his capacity for self-transcendence, and of God in respect of His concern to proceed by way of the historical and incarnational; hence the final caveat, testifying, not to a weakness of political will, but to the perennial problem as Dante himself came to sense it of reconciling the exigencies of an Aristotelian and of a Christian conscience.

(iv)

The letters, the De situ, and the eclogues

Dante, to go by what Boccaccio says, must have been a prolific correspondent: 'Gifted poet as he was,' he says in his life of Dante, 'he wrote a good number of prose letters in Latin, many still surviving.'[26] The chronicler Villani goes even further. Speaking of the three great letters addressed to the Florentines, to Henry of Luxemburg, and to the Italian cardinals (vi, vii, and xi), he refers to their exalted and authoritative style: 'All in elevated Latin and replete with wisdom and learning, they were much commended by those expert in such things' (*Cronaca* ix. 136). We have touched on their circumstances in introducing the *Monarchia*. Henry VII, count of Luxemburg, had been elected king of the Romans in 1308, declaring it his purpose to resurrect imperial power in Italy and to restore peace and unity to the empire as a whole. Clement V, pope at the time, was all in favour. Negotiations were completed and a coronation date agreed, whereupon Henry prepared to cross the Alps, which, to an enthusiastic welcome from the princes and prelates awaiting him on the other side, he did in 1310. This is the occasion of Dante's letter to the 'princes, senators of the sacred city, dukes, marquises, counts, and peoples generally of Italy' (v), composed probably towards the end of that year, in October or November. Rapturously anticipatory in tone, it calls upon the Italian people to embrace the emperor-elect and to welcome him as a new political saviour. His wrath will be great but his mercy infinite. Let none be afraid, but let all recognize the divine plan now made manifest, the long-awaited restoration of dual papal and imperial authority to Rome.

But all did not go according to plan. Under-equipped and ignorant of the complexities of Italian political thinking, Henry let slip one opportunity after another. Rebellion broke out everywhere, and much the best part of 1311 was wasted in laying siege to Brescia. This was the occasion of Dante's letters to Henry (April 1311; VII) and to the 'iniquitous Florentines within' ('scelestissimis Florentinis intrinsecis'; March 1311; VI). The letter to Henry sounds a note of great urgency. Never mind Milan, Brescia, Cremona, and the other seats of rebellion in the north, Dante says, for these are simply the hydra's heads. The real culprit is on the Arno: Florence—she is the guilty party, the architect of all subversion. Put her down, and progress will be swift. The Florentines for their part are severely castigated. Do they not realize, Dante says, that the empire is instituted by God, and that in their contumacy they are laying themselves open to His wrath? Do they for one moment imagine that they can stand against the might of the emperor? What liberty can there be in partisanship? Freedom lies, not in rebellion, but in service to the law. Even so, Dante concludes, it is not too late. Repent, and all will be well. Resist, and perish.

Florence, in the event, resisted. Or rather, she was not put to the test. Henry decided to make directly for Rome, with a view to settling his account with the dissident Guelphs at a later stage. But, as we saw earlier, his arrival and coronation in Rome was a fiasco, all the more so in that, by calling Robert of Anjou to book for impeding the imperial progress, Henry damaged beyond repair his relations with Clement. It was all over. Once crowned, Henry made his way north with a view to punishing the Tuscans, but died of fever in the Pisan marshes in August 1313. Clement died the next year, in April 1314.

The next step, therefore, was to elect a new pope. Duly, the cardinals went into conclave at Avignon to choose a successor, but the conclave was bitterly divided, taking another two years to come to any decision. Eventually, in August 1316, it settled on a Frenchman from Cahors who took the name John XXII. But predating the election, and attempting to sway it in the direction of an Italian rather than of a French solution, is Dante's letter of May or June 1314 to the Italian cardinals (XI). This too is a large-scale piece, broadly conceived and amply delivered. At its heart are the twin notions of a decadent priesthood, time-serving in its political manoeuvring, and of a destitute Rome, deprived now both of its emperor and of its pope. Let the clergy take warning, Dante says, from the Nadab and Abihu episode in Leviticus 10—they offered strange fire to the Lord and were themselves devoured by fire—and from the buying and selling in the temple episode in John 2—Christ drove the dealers out with a whip. The Church is moribund. She is being dragged by her leaders into the wilderness, into lasting perdition—and what wonder, Dante goes on, if the great luminaries of the Church (Gregory, Ambrose, Augustine, the Damascene, and Bede) are all passed over in favour of the decretalists (latter-day commentators on the papal decrees or decretals) and others of shallow wit and learning? No, Dante says, let the Italian cardinals reflect again on the unhappy condition of Rome, seat both of the empire and of the papacy, and unite in putting down the wretched Gascons once and for all. In the event, however, a Gascon was elected, and Christendom was plunged within little more than a decade into the bitterest kind of moral and political conflict.

It is in this context, then, that Dante develops the leading ideas of his political programme, most of them now familiar to us from the *Monarchia*. First comes the all-important notion of authority flowing in parallel from God, of the dual appointment of pope and emperor directly; so, for example, the 'Him from whom, as from a point, the power of Peter and of Caesar bifurcates' clause of v. 5, or the 'gracious providence' passage from vi. 1:

> The gracious providence of the Eternal King, who in His goodness ever rules the affairs of the world above, yet ceases not to look down on our concerns here below, committed to the Holy Roman Empire the governance of human affairs, to the end that mankind might repose in the peace of so powerful a protection, and everywhere, as nature demands, might live as citizens of an ordered world.

Then comes the providentialism of Rome, the sanctioning of her ascendancy by God, and, more strikingly still, the christological status of the emperor, a Christ-like saviour-figure:

> . . . the triumphant Henry, elect of God, thirsting not for his own but for the public good, has for our sakes undertaken his heavy task, sharing our pains of his own free will, as though to him, after Christ, the prophet Isaiah had pointed the finger of prophecy, when by the revelation of the Spirit of God he declared, 'Surely he hath borne our griefs and carried our sorrows.'
>
> (vi. 6)

Prominent too, in the letter to the iniquitous Florentines, is the notion of obedience to the law as freedom ('observance whereof, if it be joyous, if it be free, is not only no servitude, but to him who observes with understanding is manifestly in itself the most perfect liberty', vi. 5), as, in xi, is Dante's indictment of the Church as negligent of its true wealth:

> And what wonder? Each one has avarice to wife, even as you yourselves have done; avarice, the mother never of piety and of righteousness, but ever of impiety and unrighteousness. Ah! most loving Mother, Spouse of Christ, that by water and the spirit bearest sons unto thy shame! Not charity, not Astraea, but the daughters of the horseleech have become thy daughters-in-law. And what offspring they bear thee all but the Bishop of Luni bear witness. Your Gregory lies among the cobwebs; Ambrose lies forgotten in the cupboards of the clergy, and Augustine with him; and Dionysius, the Damascene, and Bede; and they cry up instead I know not what *Speculum*, and Innocent, and him of Ostia. And why not? Those sought after God as

their highest end and good; these get for themselves riches and benefices.

(XI. 7)

Resplendent as these ideas are, however, as well as decisive for the shape of the *Monarchia* to come, their lasting interest lies in the way in which, like *Tre donne* and *Doglia mi reca*, they witness to the experience of exile. True, ideology looms large, and these, ostensibly, are essays in God's grand design. But informing and authorizing them from within is the yearning, the anger, the sense of dislocation, and the pleading of one himself displaced and far from home; hence their value, not simply, nor even primarily, as political blueprint, but as testimony yet again to the suffering spirit, to the awareness and sensibility of one living out the substance of his discourse.

Other letters too relate to politics. No. II, for instance (probably early 1304), addressed to the Counts Oberto and Guido da Romena, laments the death of their uncle Alessandro, fêted here—presumably before Dante came into possession of the information determining the contrary verdict in the *Inferno*—as a paragon of virtue. Here too, though, the most interesting aspect of the letter is the subjective-exilic aspect, that part of it relating to Dante's own predicament: 'Poverty, like a vindictive fury, has thrust me deprived of horses and arms into her prison den, where she has set herself relentlessly to keep me in durance; and though I struggle with all my strength to get free, she has hitherto prevailed against me' (II. 3)—lines echoing the 'Ahi, piaciuto fosse al Dispensatore de l'universo' passage of *Convivio* I. iii. 3–5. Political too, in so far as they relate to the well-being of the Emperor, are the so-called Battifolle letters (VIII, IX, and X), letters of good wishes and of kindly reciprocation dating from 1311 and written to the Empress by Dante on behalf of, and in the name of, the Countess Battifolle (unable, presumably, to do the honours herself). More interesting, however, and indeed of importance for the reconstruction of Dante's experience in exile, are I and XII, the former addressed to Cardinal Niccolò da Prato and belonging to the early stages of Dante's exile (1304), and the latter addressed to a friend in Florence and belonging to the later stages (1315). The Niccolò da Prato letter predates Dante's split with his White Guelph colleagues, for, speaking from Arezzo on behalf of them all, it confirms the White Guelph intention to submit peaceably to the conciliatory mission of the Cardinal to Florence in the spring of 1304, peace and freedom being their only concern. 'What else did our White standards seek,' Dante asks, 'and for what else were our swords and our spears dyed with crimson, save that they, who at their own mad will and pleasure have maimed the body of civil right, should submit their necks to the yoke of beneficent law, and should be brought by force to the observance of their country's peace?' (I. 2) This, then, is a moment of idealism and of anticipation, of faith in the ultimate triumph of justice and of hope for a speedy home-coming. The letter to a friend in Florence, however, is quite different, for here, though home-coming remains a possibility, it is home-coming on terms Dante deems humiliating. Who the friend in Florence was is not certain, possibly Dante's brother-in-law Teruccio di Manetto Donati (to go by the letter's 'Pater', very likely a priest). But whoever he was, he had evidently written in an affectionate manner informing Dante of the conditions settled on for his return, conditions, however, involving not only a solemn procession to, and sponsored reconsecration at, the Baptistery but the wearing of a suitably inscribed mitre. Dante is scandalized, and angrily refuses to concur. Is he, he asks, a common wretch thus to be paraded publicly? No, he says, this is not the way. To return with the honour befitting a philosopher and man of learning, fine. Anything less than that, however, and he will not return at all. This, then, represents the final stage in Dante's long and complex relationship with his native city, and the closing lines of his letter, among the most moving in all his Latin writings, are worth quoting in full:

No, my father, not by this path will I return to my native city. If some other can be found, in the first place by yourself and thereafter by others, which does not derogate from the fame and honour of Dante, that will I tread with no lagging steps. But if by no such path Florence may be entered, then will I enter Florence never. What! can I not anywhere gaze on the face of the sun and the stars? Can I not under any sky contemplate the most precious truths, unless I first return to Florence, disgraced, nay dishonoured, in the eyes of my fellow citizens? Assuredly bread will not fail me!

(XII. 4)

Passing over the letters to Cino da Pistoia and to Moroello Malaspina, letters we have already touched on in connection with Dante's post-exilic *rime*, we come now to—the political letters apart—the most important item in the collection, the letter to Can Grande della Scala (XIII) dating probably from about 1319. And here straight away we are in difficulty; for though until fairly recently the authenticity of the letter had by and large been accepted on the basis of a long and remarkable study by Edward Moore (1903),[27] now, however, and as a result of enquiries conducted on more specifically stylistic grounds (on the basis of Dante's use of the *cursus* or Latin rhythmic cadence, a feature of medieval epistolary style generally), the whole question is once more *sub judice*.

The letter is in three parts: an epistolary part proper (paragraphs 1–4), a methodological part (paragraphs 5–16), and an expository part (paragraphs 17 to the end). The first part, the object of which is to confirm Can Grande as the dedicatee of the *Paradiso*, is a kind of extended *captatio benevolentiae*, a statement of the author's affection and esteem. Lodged probably with Guido da Polenta at Ravenna, but anxious to meet again the great Ghibelline leader of Verona, Dante explains how his first visit there surpassed every expectation. Can Grande was every inch the noble and virtuous prince he had heard about, the ground, he goes on to explain, of true friendship between them. Never mind, he says, their *formal* inequality (Can Grande is a prince and he a pauper); what matters is the communion of souls, the unity of illustrious spirits sealed now in the gift of the *Paradiso*: 'And I have found nothing more suitable even for your exalted station than the sublime canticle of the *Comedy* which is adorned with the title *Paradise*: this, then, dedicated to yourself, with the present letter to serve as its superscription, I inscribe, offer, and in short commend to you' (XIII. 3). With this the epistolary section ends and the methodological part begins, covering as it goes on such questions as the *Commedia*'s overall shape, structure, and discursive mode (its *forma tractatus* and *forma tractandi*), the philosophical genre to which it belongs (moral philosophy), the significance of the term 'comoedia' (a story with a happy ending), and the nature of the poem's allegory. Of these, the most important and far-reaching in its implications is the last, the allegorical section,[28] the key paragraphs, 7 and 8, running as follows:

7. For the elucidation, therefore, of what we have to say, it must be understood that the meaning of this work is not of one kind only; rather the work may be described as 'polysemous' [*polysemos*], that is, having several meanings; for the first meaning is that which is conveyed by the letter, and the next is conveyed by what the letter signifies—the former of which is called the literal sense, while the latter is called the allegorical or the moral or the anagogical sense. And for the better illustration of this method of exposition we may apply it to the following verses: 'When Israel went out of Egypt, the house of Jacob from a people of strange language; Judah was his sanctuary, and Israel his dominion.' For if we consider the letter alone, the thing signified to us is the going out of the children of Israel from Egypt in the time of Moses; if the allegory, our redemption through Christ is signified; if the moral sense, the conversion of the soul from the sorrow and misery of sin to a state of grace is signified; if the anagogical, the passing of the sanctified soul from the bondage of the corruption of this world to the liberty of everlasting glory is signified. And although these mystical meanings are called by various names, they may one and all in a general sense be termed allegorical, inasmuch as they are different from the literal or historical; for the word 'allegory' is so called from the Greek *alleon* which is the Latin *alienum* or *diversum*.

8. This being understood, it is clear that the subject in regard to which these alternative meanings operate must itself be twofold. The subject of this work must therefore be considered (*a*) from the point of view of its literal meaning, and (*b*) from that of its allegorical interpretation. The subject, then, of the whole work, taken in the literal sense only, is the state of souls after death pure and simple; for on and about that the argument of the whole work turns. If, however, the work be regarded from the allegorical point of view, the subject is man according as by his merits or demerits in the exercise of his free will he is deserving of reward or punishment by justice.

The importance of these paragraphs lies in their bearing on the essentially 'figural' structure of the *Commedia* and on how, therefore, it is to be read and interpreted. The question is complicated, but the gist of it is that while allegory proper involves the notion of saying one thing and meaning another ('aliud dicitur aliud intellegitur' was the classic formula),[29] figuralism involves the idea of one event or character anticipating, illuminating, and in turn deriving further significance from another (so, to use Dante's example, the Old Testament exodus from Egypt both anticipates and, in respect of its inner meaning, is fulfilled by man's exodus in Christ from the bondage of sin). Now Dante, it is true, refers in the lines just quoted to the *difference* between the various meanings of his text; the allegorical (in the strict Christological sense of the term), the moral, and the anagogical, he says, differ from the historical or literal ('quum sint a literali sive historiali diversi'); but difference is incorporated in a pattern of thought involving, not the substitution of meanings, but their mutual invocation and indeed mutual inherence. Thus the literal meaning of the poem identified in paragraph 8 ('the state of souls after death pure and simple') does not stand in for the allegorical meaning ('man as deserving of reward or punishment through the exercise of free will'), taking its place until such time as the penny has dropped and it is no longer needed; rather, it takes up and confirms the allegorical meaning in its ultimately eschatological truth. It seals it prophetically in its—short of an act of will this side of death—inner inevitability. With this, needless to say, allegory as a way of reading and appreciating the *Commedia* is superseded, and though allegorical elements remain in the poem (the three beasts in *Inferno* I, for example, or the pageantry of the earthly paradise towards the end of the *Purgatorio*), they are absorbed in a tendency to signify, not alternatively, but simultaneously, in a single imaginative act.

Worth noting finally, as far as the Epistle to Can Grande is concerned, is the expository part of the letter beginning in paragraph 17. In fact, the detailed exposition does not get very far; or, rather, lines 1–36 of *Paradiso* I are dealt with exhaustively in paragraphs 17–32, the rest of the canticle being passed over in the broad sweep of paragraph 33. No matter, though, since what Dante has to say about the 'prologue' of the canticle, the 'La gloria di colui che tutto move' passage beginning at line 1, confirms splendidly the Neoplatonist or idealist aspect of his theological stance in the *Paradiso* as a whole. Having nodded, then, in the direction of Aristotle and of the 'impossibility of infinite regression' argument—there must, he says, be a first cause or else you would go back for ever (paragraph 20)—Dante develops his exposition in Pseudo-Dionysian categories. Everything that *is*, he says, is an out-shining of the primal light, the Intelligences reflecting like mirrors the light received from above to those below, a pattern of thought bearing implications both on the linguistic and on the psychological front (28 and 29). Linguistically it is a question of ineffability, of the need in such a context to resort to metaphor and to other oblique forms of signification; while psychologically it is a question of the mind so losing itself in the divine that it is stilled by its own self-transcendence:

> 28. And after he [the author] has said that he was in the Paradise described by him by means of a circumlocution, he goes on to say that he saw certain things which he who descends therefrom is powerless to relate. And he gives the reason, saying that 'the intellect plunges to such depth' in its longing, which is for God, 'that the memory cannot follow'. For the understanding of which it must be noted that the human intellect in this life, by reason of its connaturality and affinity to the separate intellectual substances, when in exaltation, reaches such a height of rapture that after its return to itself memory fails, since it has transcended the range of the human faculty.

Whether all this implies mystic experience in Dante himself is a moot point; not necessarily, for the psychology of ecstasy had been explored for all to contemplate by Augustine, Bernard, and—especially—Richard of St Victor (all duly acknowledged in paragraph 28).[30] What matters about this section of the letter is the way in which it confirms the presence of the Augustinian, of the Benedictine, and of the Victorine as spiritual possibility in Dante, as one of the ways in which, provisionally at least, the surpassing truth of God might be known and celebrated here below.

The final two works we have to consider, the *Quaestio de situ et forma aque et terre* and the eclogues, may conveniently be taken together, for they each represent a form of self-justification, a sort of *apologia pro vita sua*. Dante, nearing the end of his activity as a poet-theologian in the vernacular, but feeling around him a shift in taste and literary fashion, moves on to the defensive. The *De situ*, a learned disquisition or *quodlibet* delivered before the *clerus* or intellectual élite of Verona in January 1320, has, as the title suggests, to do with the earth and the sea and how they relate to one another. The point is that, officially, the cosmos as a whole was characterized by perfect concentricity. Like wheels within wheels, one sphere nested neatly within the next. First came the Empyrean, the all-encompassing First Mind, then the *primum mobile*, then the firmament or starry heaven, then the seven planetary spheres, and finally—all contained within the sphere of the moon—those of fire, air, water, and earth. But when we look at the world, we find that the earth sticks up above the water, thus compromising the ideal pattern. What are we to make of this?

The question had received a good airing, and various explanations had been forthcoming. Some maintained that it was a matter of eccentricity on the part of the spheres, that they were not, after all, contained neatly one within the other. But that, Dante says, will not do, for it would imply a double centre of gravity in the universe, earth and water each tending to the centre of its own sphere. Others had favoured a system of humps and bumps (*gibbositas*), which, while preserving a single centre of gravity for both earth and water, would none the less account for the prominence of dry land in the inhabited part of the world. But neither will this do, for humps on the sphere of the earth would—if concentricity is to be maintained—necessitate humps on the sphere of the water, which is manifestly impossible (water tending always to settle uniformly). So how are we to account for what looks like a gross irregularity?

The answer, Dante explains (paragraph 18), lies in the purposes of nature; for while she wishes the universe to be concentric, she also wills the realization of every complex potentiality, which means not so much separating as mixing the elements at key points in the whole:

> 19. Although, therefore, the earth in its simplicity seeks equally the same central point . . . even so there is that in its nature which, according to Nature's law generally, suffers it in part to be raised in order that a mingling might be possible.

That, then, is that; it is all a question of earth and water combining in keeping with nature's plan. And it is the 'that's that'—Dante's triumphant sense of settlement—that counts; for Dante was a theological poet, and a theological poet, moreover, in the vernacular. And as such he had, at an earlier stage, offered a vernacular-theological solution to this very question. The passage concerned, *Inferno* XXXIV. 121–6, runs as follows:

> 'Da questa parte cadde giù dal cielo;
> e la terra, che pria di qua si sporse,
> per paura di lui fé del mar velo,
>   e venne a l'emisperio nostro; e forse
> per fuggir lui lasciò qui loco voto
> quella ch'appar di qua, e sù ricorse.'

The explanation is nothing if not theological. Originally, Dante is saying, the earth stood proud of the sea in the southern hemisphere. But with Lucifer's expulsion from paradise it took flight, leaving behind it—Mount Purgatory apart (formed by a simultaneous voiding of the ground beneath Jerusalem)—only the 'veil' of the sea. From then on there was dry land in the northern hemisphere, in the 180 degrees between the Ganges and Cadiz comprising the known and inhabited world. The philosophers, however, were sceptical. They were sceptical both of the explanation, and of Dante its author. Who was he, a mere verse-maker in the vernacular, to pronounce thus philosophically? Theologically maybe; but philosophically, surely not. And that, fundamentally, is what the *Quaestio* is about. It sets out to insist (*a*) that there is more in heaven and earth than the philosophers have dreamt of, and (*b*) that Dante, when so minded, is as capable as the next man of doing philosophy, of producing reasonable answers. The relevant passage as far as (*a*) is concerned comes in paragraph 22:

> 22. Let men desist, therefore, let them desist from searching out things that are above them, seeking instead, inasmuch as they are able, 'to be drawn up as far as may be to things immortal and divine', leaving aside whatsoever is too great for them. Let them listen to the friend of Job when he says, 'Will you understand the footprints of God, and search out the Almighty to perfection?' Let them listen to the Psalmist when he says, 'Thy knowledge is wonderful and has comforted me, and I may not attain to it.' Let them listen to Isaiah when he says, 'As far as the heavens are above the earth, so far are my ways above your ways' . . . And let this suffice for the enquiry into the truth we have undertaken to look into.

while decisive in respect of (*b*) is the following from the end of the treatise:

24. This philosophical demonstration was accomplished by me, Dante Alighieri, least among philosophers, in the reign of the all-victorious lord Can Grande della Scala, lord in the name of the Holy Roman Empire, in the noble city of Verona in the sanctuary of the glorious Helen, before the gathered clergy of Verona—with the exception of some who, burning with a superabundance of love, decline the invitation of others, refusing, as mendicants through humility of the Holy Spirit, and lest the excellence of others appear to be endorsed, to be present at their discourses.

The tone is searing, irony gradually shading off into the most virulent sarcasm, this in turn confirming Dante's aim in the piece as a whole—to settle his score with those who felt that, philosophically, he was a non-starter, a mere fanciful theologizer.

Spirited too, but cleverer, is Dante's exchange with Giovanni del Virgilio, a Bolognese professor of classics expert on the great poet of Latin antiquity.[31] There are two items to consider, the first, the eclogue beginning *Vidimus in nigris albo patiente lituris* ('We saw in black letters grounded on white'), being the wittier and more accomplished. Giovanni del Virgilio, representative of the new pre-Petrarchan brand of literary humanism now gaining ground in the north of Italy, had sent Dante a metrical epistle praising him for the lofty substance of his poem so far, but lamenting his failure to produce something good and literary in Latin. How about an epic on the exploits of, say, Henry VII or Can Grande della Scala, a certainty, he says, for the laurel crown?

> En ego jam primus si dignum duxeris esse
> clericus Aonidum vocalis verna Maronis
> promere gymnasiis te delectabor ovantum,
> inclita Peneis redolentem tempora sertis;
> ut præfectus equo sibi plaudit preco sonorus
> festa trophœa ducis populo praetendere læto.

Lo I, taking the lead if thou deem me worthy, clerk of the Aonides, vocal attendant upon Maro, shall rejoice to present thee to the applauding schools, thine illustrious temples redolent with the Peneian garlands; even as the herald mounted on his steed exults to proclaim, with ringing voice, to the rejoicing people the festal triumphs of the General.

(Giovanni del Virgilio, I. 35–40)

The style, clearly, is out of the ordinary. It is the style of the pastoral eclogue, originally Greek but by the time of Virgil a vehicle for moral, social, political, prophetic, and satirical commentary on current affairs. And this, the witty obliqueness of the eclogue, is something Dante evidently relished, for, quite as ably, he replied in kind. Giovanni del Virgilio's missive, he says, arrived just as he and his friend Melibœus were tending their goats on the hillside (i.e. teaching rhetoric, probably, in Ravenna). Melibœus urges Dante (Tityrus in the poem) to divulge the contents of the letter, which he eventually does. Mopsus (Giovanni del Virgilio), alone among so many lawyers in his dedication to the muses, has invited him, he says, to the utopia Menalus (Bologna), there to exercise his art in more civilized circumstances. 'You should accept,' replies Melibœus, 'for otherwise you will be a rustic shepherd for the rest of your days.' 'No,' says Tityrus, 'I shall await my laurel crown on the strength of my accustomed song. In the mean time, and as a sign both of contentment and of respect, I'll send ten pailfuls of milk—cantos, no doubt, of the *Paradiso*:[32]

> 'Est mecum quam noscis ovis gratissima' dixi,
> 'ubera vix quae ferre potest, tam lactis abundans.
> Rupe sub ingenti carptas modo ruminat herbas.
> Nulli juncta gregi nullis assuetaque caulis
> sponte venire solet nunquam vi poscere mulctram.
> Hanc ego præstolor manibus mulgere paratis;
> hac implebo decem missurus vascula Mopso.'

'I have', said I, 'one sheep, thou knowest, most loved; so full of milk she scarce may bear her udders; even now under a mighty rock she chews the late cropped grass; associate with no flock, familiar with no pen, of her own will she ever comes, ne'er must be driven to the milking pail. Her do I think to milk with ready hands; from her ten measures I will fill and send to Mopsus.'

(Dante, I. 58–64)

This occasioned a further round of correspondence in the pastoral mode, Giovanni del Virgilio ever more fervent in his praise of Dante the new Virgil, and Dante for his part ever more fearful of the Polyphemus-like ogre (Guelph reprisals, perhaps, for a staunch critic of the party) awaiting him should he dare set foot in Bologna. Again, the tone is allusive and clever. And that—cleverness—is the point; for Dante was no fool. He knew exactly what was going on and what men were saying about him—that as an old-style scholasticizing visionary in the vernacular he had had his day. Deftly, then, but with an assertiveness as fierce as anything in his work, he commits himself in mid-bucolic—in mid-mastery of the *new*—to the *rationes* of the *old*, to the old vernacular-prophetic way. This, he believes, will win him the laurel crown and the glorious homecoming he so deeply craves. In the event, he was mistaken, for the poem which he thought might win him the highest poetic honours did not. They went instead to the neo-Senecan tragedian Albertino Mussato, a good man but no Dante. But in another respect—in his sense of the impossibility of translating what he was and what he had to say into the categories of Renaissance humanism—he was wholly right.

Conclusion

THERE remains only one question to ask: in what sense are Dante's minor works minor? The answer to that depends on how we look at them. Looked at diachronically—sequentially, that is, or in respect of what happened next—there is no doubt that they lack the *Commedia*'s deep sense of unity in diversity, its coalescence of the many and varied in the one. Typically, everything stands over and against everything else, jostling with it, even, in the bid for priority; so, for example, the *Convivio*, with its mapping out of the human and the more-than-human, of the here and now and the hereafter; or the *De vulgari eloquentia*, with its distinction between Italian and the less than Italian, between the illustrious and the municipal; or the *Monarchia*, with its separating out of the earthly paradise and the heavenly paradise, of philosophy and theology, of present and future possibility. Everything is contradistinction, reclamation, vindication. True, even in the *Commedia* distinctions remain; but underlying and transcending them is a fresh sense of continuity, of the way in which diversity stands ultimately to be taken up in unity, in a totality of being, seeing, and understanding acknowledging, certainly, but subsuming difference. Thus the theological comprehends the philosophical, the hereafter confirms the here and now, the heavenly completes the earthly, grace informs nature, and the Christian renews the pagan. None of this is envisaged in the minor works, which diachronically, therefore, must be regarded as superseded by the major work, as overtaken by a new literary, linguistic, and philosophical maturity.

But if diachronically the minor works are overtaken by the major work, synchronically, or from the point of view of what each represented *there and then*, in the moment of its composition, the picture is different; for what each represented there and then was a moment of superlative self-awareness, of unclouded insight into the reasons of the author's own being. Take, for instance, the *Vita nuova*. This, as we have seen, is self-consciously provisional in conception, disposed as of the essence to resolution on a more elevated plane (the 'studio quanto posso' passage of XLII. 2). But for all its provisionality, it remains from the synchronic point of view, from the 'there and then' point of view, total and incontrovertible, a faithful register of present possibility and of current commitment. And the same goes for the *Convivio* and for *De vulgari eloquentia*; philosophically precarious and beset by all kinds of difficulty, each, as confession, is complete, utterly faithful to every inflection of Dante's mood and thought. So it is with all his works, major and minor alike; they are all in the nature of apology, of an essay in self-intelligibility. Each, whatever its ostensible purpose in propounding a theology or metaphysic, in righting a social or political wrong, or in setting out the principles of good literary practice, seeks to establish the rationale of Dante's own being and doing, the intimate structure of his own humanity. Here, then, in their role as witness, in their taking seriously 'the Augustinian notion of the text as confession', lies their power and importance. Deny them their confessionalism, and they are nothing, a mere accretion of scholastic lore and mystification. Grant it, and they shine scarcely less brightly than the *Commedia*.

Abbreviations

AGI	*Archivio glottologico italiano*
AR	*Archivio Romanicum*
ARCH	*L'archiginnasio*
ASNSP	*Annali della Scuola Normale di Pisa*
BCSFLS	*Bollettino del Centro di studi filologici e linguistici siciliani*
BEL	*Belfagor*
BSDI	*Bollettino della Società Dantesca*
CN	*Cultura neolatina*
CO	*Convivium*
CS	*Cultura e Scuola*
DDJ	*Deutsches Dante-Jahrbuch*
DS	*Dante Studies*
ED	*Enciclopedia dantesca*
FR	*Filologia romanza*
GD	*Giornale dantesco*
GSLI	*Giornale storico della letteratura italiana*
IMU	*Italia medioevale e umanistica*
IT	*Italica*
ITST	*Italian Studies*
LC	*Letture Classensi*
LM	*Letterature moderne*
LS	*Lingua e stile*
MAEV	*Medium Aevum*
MGH	*Monumenta Germaniae Historica*
MH	*Medievalia et Humanistica*
MLN	*Modern Language Notes*
MLR	*Modern Language Review*
MP	*Modern Philology*
MR	*Medioevo romanzo*
MS	*Medieval Studies*
NLD	*Nuove Letture Dantesche*
NSM	*Nuovi studi medievali*
RANL	*Rendiconti dell'Accademia Nazionale dei Lincei*
RCCM	*Rivista di cultura classica e medievale*
RF	*Romanische Forschungen*
RDF	*Rassegna di filosofia*
RFNS	*Rivista di filosofia neo-scolastica*
RMAL	*Revue du Moyen Age Latin*
RP	*Romance Philology*

RR	*Romanic Review*
RSF	*Rivista di storia della filosofia*
SD	*Studi danteschi*
SFI	*Studi di filologia italiana*
SFRAN	*Studi francescani*
SM	*Studi medievali*
SP	*Speculum*
SR	*Studi romanzi*
STPH	*Studies in Philology*
TR	*Traditio*
VR	*Vox Romanica*
ZRP	*Zeitschrift für Romanische Philologie*

Notes

PREFACE

1. 'Dante's Canzone *Tre donne*' in *God's Tree*, London, 1957, 15. On the theological problems of the *Convivio*, idem, 'The Two Dantes', in the volume of that name, London, 1977, 156-253 *passim*.

CHAPTER 1

1. On 'courtly love': C. S. Lewis, *Allegory of Love. A Study in Medieval Tradition*, Oxford, 1936; E. Gilson, 'Saint Bernard and Courtly Love', in *The Mystical Theology of St Bernard*, app. iv, London, 1940; M. Valency, *In Praise of Love. An Introduction to the Love Poetry of the Renaissance*, New York, 1958; D. W. Robertson Jr., 'Some Medieval Doctrines of Love', in *A Preface to Chaucer*, Princeton UP, 1962, 391-503; A. Scaglione, *Nature and Love in the Late Middle Ages*, Berkeley, 1963; K. Foster, 'Courtly Love and Christianity', Aquinas Paper 39, London, 1963, now in *The Two Dantes*, London, 1977, 15-36; P. Dronke, *Medieval Latin and the Rise of the European Love Lyric*, 2 vols., Oxford, 1965-6; F. X. Newman (ed.), *The Meaning of Courtly Love*, New York State UP, 1968; D. Kelly, 'Courtly Love in Perspective. The Hierarchy of Love in Andreas Capellanus', *TR* 24 (1968), 119-48; F. L. Utley, 'Must We Abandon the Concept of Courtly Love?', *MH* NS 3 (1972), 299-324; E. Salter, 'Courts and Courtly Love', in *The Mediaeval World* (ed. D. Daiches and A. Thorlby), London, 1973, 407-84; J. Frappier, *Amour courtois et table ronde*, Geneva, 1973; F. Heer, 'Courtly Love and Courtly Literature', in *The Medieval World. Europe 1100-1350*, London, 1974, 153-92; J. Ferrante et al., *In Pursuit of Perfection. Courtly Love in Medieval Literature*, New York, 1975; R. Boase, *The Origin and Meaning of Courtly Love*, Manchester UP, 1977.
2. On Ovid in the Middle Ages: S. Battaglia, 'La tradizione di Ovidio nel medioevo', *FR* 6 (1959), 185-224; F. Munari, *Ovid im Mittelalter*, Zurich, 1960; *Ovid Renewed: Ovidian Influence on Literature and Art from the Middle Ages to the Twentieth Century*, ed. C. Martindale, CUP, 1988. On Ovid and 'courtly love': D. Scheludko, 'Ovid und die Trobadors', *ZRP* 54 (1934), 128-74; N. Shapiro, *The Comedy of Eros. Medieval French Guides to the Art of*

Love, Chicago, 1971; R. Boase, *The Origin and Meaning of Courtly Love*, Manchester UP, 1977.
3. Ed. S. Battaglia, Rome, 1947 (with facing early Italian); P. G. Walsh, London, 1982 (with facing English). There is an English translation, with introduction, by J. J. Parry, *The Art of Courtly Love*, Columbia UP, 1941.

In general: D. W. Robertson Jr., 'The Subject of the *De Amore* of Andreas Capellanus', *MP* 50 (1952–3), 145–61; idem, 'Some Medieval Doctrines of Love', in *A Preface to Chaucer* (n. 1 above), 391–503; L. Pollmann, *Der Tractatus* De Amore *des Andreas Capellanus und seine Stellung in der Geschichte der Amortheorie*, Freiburg, 1955; W. T. H. Jackson, 'The *De amore* of Andreas Capellanus and the Practice of Love at Court', *RR* 49 (1958), 243–51; F. Schlosser, *Andreas Capellanus. Seine Minnelehre und das christliche Weltbild des 12. Jahrhunderts*, Bonn, 1960; J. F. Benton, 'The Court of Champagne as a Literary Center', *SP* 36 (1961), 551–91; idem, 'The Evidence for Andreas Capellanus Re-examined Again', *STPH* 59 (1962), 471–8.
4. Editions and anthologies: B. Panvini, *Le rime della scuola siciliana*, i (introduction, text, and apparatus), Florence, 1962; ii (glossary), Florence, 1964; G. Contini, *Poeti del Duecento*, (2 vols.), Milan–Naples, 1960 (definitive for those poets included); other collections, among the many available, Carlo Salinari, *La poesia lirica del Duecento*, Turin, 1951; A. Giuliani, *Antologia della poesia italiana dalle origini al Trecento* i, Milan, 1975. For Giacomo da Lentini, *Poesie*, ed. R. Antonelli, Rome, 1979. A selection with translation in F. Jensen, *The Poetry of the Sicilian School*, New York and London, 1986.

On the Hohenstaufen: G. Masson, *Frederick II of Hohenstaufen*, London, 1957; T. C. Van Cleve, *The Emperor Frederick II of Hohenstaufen: Immutator Mundi*, Oxford, 1972.

Criticism: S. Santangelo, 'La scuola poetica siciliana del secolo XIII', *SM* 17 (1951), 21–45; A. Schiaffini, 'La prima elaborazione della forma poetica italiana', in *Momenti di storia della lingua italiana*, 2nd edn., Rome, 1953, 7–42; A. Monteverdi, *Studi e saggi sulla letteratura italiana dei primi secoli*, Milan–Naples, 1954 (various items); R. Baehr, 'Die Sizilianische Dichterschule und Friedrich II', in *Probleme um Friedrich II*, ed. J. Fleckenstein, Sigmaringen, 1974, 93–107 (with bibliography); E. Pasquini and A. E. Quaglio, *Le origini e la scuola siciliana*, 2nd edn., Bari, 1975 (with bibliography).

On language, style, and metrics: E. H. Wilkins, 'The Invention of the Sonnet', *MP* 13 (1915), 436–94 (now in the *The Invention of the Sonnet and other Studies in Italian Literature*, Rome, 1959, 11–39); S. Santangelo, *Il volgare illustre e la poesia siciliana del secolo XIII*, Palermo, 1924; idem, *Il primato linguistico dei Siciliani*, Palermo, 1938; M. Corti, *Studi sulla sintassi della lingua poetica avanti lo Stilnovo*, Florence, 1953; E. Li Gotti, *La questione dei 'Siciliani'*, Palermo, 1954; W. Th. Elwert, 'Per una valutazione stilistica dell'elemento provenzale nel linguaggio della scuola poetica siciliana', in *Homenaje a Fritz Krüger*, Mendoza, 1954, ii. 85–111; P. R. J. Hainsworth, 'Artifice in "Pir meu cori alligrari"', *ITST* 29 (1974), 12–27; R. R. Vanasco, *La poesia di Giacomo Lentini: analisi strutturale*, Bologna, 1979; C. Kleinhenz, *The Early Italian Sonnet: the First Century (1220–1321)*, Lecce, 1986.

On Dante and the Sicilians (in addition to introductions to the *DVE*, especially Marigo and Mengaldo): B. Panvini, 'L'esperienza dei Siciliani e il volgare illustre di Dante', in *Atti del Convegno di studi su Dante e la Magna Curia*, Palermo, 1967, 236–49; M. Marti, 'Dante e i poeti della scuola siciliana' and 'Il giudizio di Dante su Guido delle Colonne', in *Con Dante fra i poeti del suo tempo*, 2nd edn., Lecce, 1971, 7–28 and 29–42.

5. Editions and anthologies: G. Contini (n. 4 above), i. 187–482 and ii. 821–33 (notes); B. Panvini, *La scuola poetica siciliana. Le canzoni dei rimatori non siciliani*, 2 vols., Florence, 1957–8; C. Salinari (n. 4 above), 199–401; A. Giuliani (n. 4 above), i. 89–162. Particular poets (Guittone apart; see next note): G. Zaccagnini and A. Parducci, *Rimatori Siculo-Toscani del Dugento*, Bari, 1915 (for Bonagiunta and various Tuscan *minori*); A. Menichetti (ed.), *Chiaro Davanzati. Rime*, Bologna, 1965; R. Bettarini, *Dante da Maiano. Rime*, Florence, 1969 (supersedes G. Bertacchi, Bergamo, 1896).

Criticism: A. Tartaro, 'Guittone e i rimatori siculo-toscani', in *Storia della letteratura italiana* (ed. E. Cecchi and N. Sapegno), Milan, 1965, i. 351–428; G. Folena, 'Cultura poetica dei primi fiorentini', *GSLI* 147 (1970), 1–42; M. Marti, *Con Dante fra i poeti del suo tempo*, 2nd edn., Lecce, 1971; E. Pasquini and A. E. Quaglio (n. 4 above). On particular poets (Guittone apart): L. Cavazza, 'Onesto degli Onesti e le sue rime', *ARCH* 29 (1934), 101–14; D. Pierantozzi, 'Bonagiunta Orbicciani campione di "trobar leu"', *CO* 1948, 873–87; A. Chiari, 'Bonagiunta da Lucca', in *Indagini e letture*, ser. ii, Florence, 1954, 8–20; G. Volpi, 'Appunti sul lessico di Monte Andrea', *AGI* 47 (1962), 143–61.

6. Editions and textual criticism: F. Egidi, *Guittone d'Arezzo. Rime*, Bari, 1940 (subject to various amendments and improvements: see especially G. Contini, *GSLI* 117 (1941), 55–82); N. Bonifazi, *Rime*

di Guittone d'Arezzo, Urbino, 1950. Selections in Contini, Salinari, Giuliani, etc. (n. 4 above).

On Guittone's circumstances: H. Wieruszowski, 'Arezzo as a Center of Learning and Letters in the Thirteenth Century', *TR* 9 (1953), 321–91.

Criticism: G. De Robertis, 'Le Rime di Guittone' in *Studi*, Florence, 1944; G. Baehr, 'Studien zur Rhetorik in den Rime Guittones von Arezzo', *ZRP* 73 (1957), 193–258 and 357–413, and 74 (1958), 163–211; M. Marti, 'Ritratto e fortuna di Guittone d'Arezzo' in *Realismo dantesco e altri studi*, Milan–Naples, 1961, 126–55; A. Tartaro, 'La conversione letteraria di Guittone', *RCCM* 7 (1965), 1057–67; C. Margueron, *Recherches sur Guittone d'Arezzo*, Paris, 1966 (comprehensive); A. E. Quaglio, 'I poeti siculo-toscani', in E. Pasquini and A. E. Quaglio (n. 4 above), especially 243–300; A. Baldi, *Guittone d'Arezzo fra impegno e poesia*, Cercola, 1975; V. Moleta, *The Early Poetry of Guittone d'Arezzo*, London, 1976.

On Guittone *prosatore*: F. Meriano (ed.), *Le lettere di Fra Guittone d'Arezzo*, Bologna, 1923; A. Schiaffini, *Tradizione e poesia nella prosa d'arte italiana dalla latinità medievale a Giovanni Boccaccio*, 2nd edn., Rome, 1943; B. Migliorini, *Storia della lingua italiana*, Florence, 1960, 137–43; C. Segre, 'La sintassi del periodo nei primi prosatori italiani' in *Lingua, stile e società*, 2nd edn., Milan, 1976, 79–270 (at 95 ff.)

Recently on Dante and Guittone: G. Bolognese, 'Dante and Guittone Revisited', *RR* 70 (1979), 172–84.

7. Anthologies and editions: M. Marti, *Poeti del Dolce stil novo*, Florence, 1969 (complete for Cino and for Frescobaldi); G. Contini (n. 4 above), ii. 443–786 and 891–917 (complete and definitive for Guinizzelli, Cavalcanti, Lapo Gianni, and Gianni Alfani); C. Salinari (n. 4 above) 26–43 and 405–616; L. Di Benedetto, *Rimatori del dolce stil nuovo*, Bari, 1939; V. Branca, *Rimatori del dolce stil nuovo*, Genoa, 1941 (reprinted Milan, 1965); A. Cordié, *Dolce stil novo*, Milan, 1942.

Criticism: B. Nardi, 'Filosofia dell'amore nei rimatori italiani del Duecento e in Dante' in *Dante e la cultura medievale*, 2nd edn., Bari, 1949, 1–92; U. Bosco, 'Il nuovo stile della poesia dugentesca secondo Dante', in *Medioevo e Rinascimento*, Florence, 1955, i. 77–101; E. Bigi, 'Genesi di un concetto storiografico: "Dolce stil novo"', *GSLI* 132 (1955), 333–71; G. Petrocchi, 'Il Dolce stil novo', in *Storia della letteratura italiana* (ed. E. Cecchi and N. Sapegno), Milan, 1965, i. 729–94; M. Marti, *Con Dante fra i poeti del suo tempo* (n. 4 above); idem, *Storia dello Stil Nuovo*, 2 vols.,

Lecce, 1973; G. Favati, *Inchiesta sul dolce stil nuovo*, Florence, 1975; A. E. Quaglio, 'Gli stilnovisti', in *Lo stilnovo e la poesia religiosa*, 2nd edn., Bari, 1975, 9–148 (with bibliography).

8. Editions: D. De Robertis, *Rime*, Turin, 1986; also, G. Favati, *Rime*, Milan–Naples, 1957; G. Contini (n. 4 above), ii. 487–567 and 899–906; idem, *Le rime di Guido Cavalcanti*, Verona, 1968; M. Ciccuto, *Rime*, Milan, 1978 (Contini's 1960 text with slight modification; introduction Maria Corti). Translation: *The poetry of Guido Cavalcanti*, ed. and trans. L. Nelson, Jr., New York, 1986.

Criticism: M. Marti, 'Arte e poesia nelle rime di Guido Cavalcanti', *CO* 1949, 178–95 (also in *Dal certo al vero*, Rome, 1962, 49–74); G. Favati, 'Tecnica ed arte nella poesia cavalcantiana', *Studi petrarcheschi* 3 (1950), 117–41; M. Corti, 'La fisionomia stilistica di Guido Cavalcanti', *RANL*, ser. viii, 5 (1950), 530–52; idem, 'Dualismo e immaginazione visiva in Guido Cavalcanti', *CO* 1951, 641–66; D. De Robertis, 'Cino e Cavalcanti o le due rive della poesia', *SM* 18 (1952), 55–107; A. E. Quaglio, 'Guido Cavalcanti' in *Lo stilnovo . . .* (n. 7 above), 88–118; C. Calenda, *Per altezza d'ingegno. Saggio su Guido Cavalcanti*, Naples, 1976.

On *Donna me prega* (text in Contini and Favati): B. Nardi, 'L'averroismo del primo amico di Dante', *SD* 25 (1940), 43–80, and 'Di un nuovo commento alla canzone del Cavalcanti sull'amore', *CN* 6–7 (1946–7), 123–35 (a radical Aristotelian or Averroist view); idem, 'Noterella polemica sull'averroismo di Guido Cavalcanti', *RDF* 3 (1954), 47–71, and 'L'amore e i medici medievali' in *Studi in onore di Angelo Monteverdi*, Modena, 1959, ii. 517–42 (especially 518–19), both in *Saggi e note di critica dantesca*, Milan–Naples, 1966, 190–219 and 238–67. Also (more generally), idem, 'Dante e Guido Cavalcanti', *GSLI* 139 (1962), 481–512 (and in *Saggi e note di critica dantesca*, 190–219), and (more generally still) idem, 'Filosofia dell'amore nei rimatori italiani e in Dante' in *Dante e la cultura medievale*, Bari, 1942, 1–88. In support of Nardi, P. O. Kristeller, 'A Philosophical Treatise from Bologna Dedicated to Guido Cavalcanti: Magister Jacobus de Pistorio and his Quaestio de Felicitate' in *Medioevo e Rinascimento: Studi in onore di Bruno Nardi*, Florence, 1955, i. 425–63. Philosophically more moderate, G. Favati, 'La canzone d'amore del Cavalcanti', *LM* 3 (1952), 422–53; idem, 'Guido Cavalcanti, Dino del Garbo e l'averroismo di Bruno Nardi', *FR* 2 (1955), 67–83; idem, *Inchiesta sul Dolce Stil Nuovo* (n. 7 above); also, J. E. Shaw, *Guido Cavalcanti's Theory of Love: The Canzone d'Amore and Other Related Problems*, Toronto,

1949 (for the Neoplatonist tradition and Albert the Great). On Cavalcanti and the commentary of Dino del Garbo: G. Favati, 'La glossa latina di Dino del Garbo a "Donna me prega" del Cavalcanti', *ASNSP* 21 (1952), 70–103; A. E. Quaglio, 'Prima fortuna della glossa garbiana a "Donna me prega" del Cavalcanti', *GSLI* 141 (1964), 336–68.

9. Edition: Contini (n. 4 above), ii. 447–85 and 893–8 (notes); complete in respect of certain attribution and supersedes Casini, Zaccagnini, and Di Benedetto. Also, *Poesie*, ed. E. Sanguineti, Milan, 1986.

    Criticism: I. Baldelli, *Poeti del 'Dolce stil novo'. Guido Guinizzelli e Lapo Gianni*, Pisa, 1963, 7–116; R. Russell, *Tre versanti della poesia stilnovistica: Guinizzelli, Cavalcanti, Dante*, Bari, 1973; M. Marti, *Storia dello Stil Nuovo* (n. 7 above); V. Moleta, ' "Come l'ausello in selva a la verdura" ', *SD* 52 (1979–80), 1–67 (reprinted in *Guinizzelli in Dante*, below); I. Bertelli, *La poesia di Guido Guinizzelli e la poetica del 'dolce stil nuovo'*, Florence, 1983.

    On Guinizzelli and Dante: V. Moleta, *Guinizzelli in Dante*, Rome, 1980; T. Barolini, *Dante's Poets. Textuality and Truth in the Comedy*, Princeton UP, 1984, especially 123–53.

10. F. Pellegrini, 'La canzone d'amore di Guido Guinizzelli', *NSM* I (1923), 119–37; L. Mascetta Caracci, 'La Canzone "Al cor gentil" di Guido Guinizzelli', *ARCH* 27 (1932), 216–32 and 344–59; F. Torraca, 'La canzone "Al cor gentil ripara sempre amore" ', *Atti della Reale Accademia di Archeologia, Lettere e Belle Arti di Napoli*, NS 13 (1933–4), 41–66; M. Casella, 'Al cor gentil repara sempre amore', *SR* 30 (1933), 5–53.

11. For the *fattore* motif cf. *Madonna, il fino amor ched eo vo porto*, ll. 37–42 (*Poeti del Duecento*, n. 4 above, ii. 454).

12. See with reference to Provençal (with bibliography) L. M. Paterson, *Troubadours and Eloquence*, OUP, 1975.

13. See too, for Cavalcanti on Guittone's faulty syllogisms, his expressive barbarity, and his general impenetrability, the sonnet *Da più a uno face un sollegismo* (*Poeti del Duecento*, ii. 557). M. Ciccuto, 'Il sonetto cavalcantiano "Da più a uno face un sollegismo" ', *Critica letteraria* 6 (1978), 305–30.

14. Contini, ed. *Rime*, x: 'Mai in lui un sospetto di scetticismo. Ci sono scherzi anche nella sua opera, ma remotissimi dai centri dell'ispirazione. In fondo, una serietà terribile . . .'

15. For Dante da Maiano, *Rime*, ed. R. Bettarini, Florence, 1969 (supersedes G. Bertacchi, Bergamo, 1896).

    On the 'Duol d'amor' *tenzone* and the problem of attribution, F. Pellegrini, 'La tenzone del "duol d'amore" tra Dante Alighieri e

Dante da Maiano', *BSDI*, NS 24 (1917), 160–8; Foster-Boyde ad loc. (ii. 6–9).

16. See—in addition to Contini, Foster-Boyde, and Barbi-Maggini (*Rime della 'Vita Nuova' e della giovinezza*, Florence, 1956) ad loc.—L. Pietrobono, 'Se la canzone di Dante "E' m'increscre di me sì duramente" si possa ritenere scritta per Beatrice', *GD* 29 (1926), 53–6. On the autobiographical aspect of *E' m'increscre di me*—its anticipation as history of the *Vita nuova*—D. De Robertis, *Il libro della Vita Nuova*, 2nd edn., Florence, 1970, 31 ff. and 82–3.

17. On *Guido i' vorrei* as parenthesis, De Robertis (n. 16 above), 81 ('. . . sonetto troppo compiaciutamente "vieux style", del resto, troppo "evasivo" per segnare un momento definito anche nella preistoria della Vita Nuova . . .'); G. Gorni, ' "Guido, i' vorrei che tu e Lapo ed io" (sul canone del Dolce Stil Novo)', *SFI* 36 (1978), 21–37.

    Lapo Gianni, in his entirety, in Contini (n. 4 above), ii. 569–603. Folgore da San Gimignano (*Sonetti de' mesi*), ibid. 403–19.

18. 'Il nuovo stile non poteva avere inaugurazione più solenne, né che meglio ne testimoniasse la carica intellettiva, l'esigenza che lo muoveva, lo stesso entusiasmo della scoperta. Per me, è la prima vera prova di forza di Dante, il primo punto fermo nella sua avventurosa carriera poetica, e al quale potremo sempre riferirci con sicurezza; nonché una conferma dell'alto destino della nostra lirica' (De Robertis (n. 16 above), 135–6). R. Crespo, 'Il proemio di "Donne ch'avete intelletto d'amore" ' in *Studi di filologia e di letteratura italiana offerti a Carlo Dionisotti*, Milan–Naples, 1973, 3–13; idem, ' "Color di perle ha quasi" e "Alcuno parlare fabuloso" ', *SD* 48 (1971), 115–19.

19. On *Tanto gentile*, G. Contini, 'Esercizio d'interpretazione sopra un sonetto di Dante', in *Un'idea di Dante: saggi danteschi*, Turin, 1976 (1947), 21–31.

20. Isidore, *Etym*. x. 243; Bonaventure, 1 *Sent*. 3. 1. 3 dub. 1, etc.

21. Fundamental for the attribution of the *Fiore* to Dante (an attribution resting on lexical, syntactic, and rhythmic elements of form common to the *Fiore* and to the canonical works), G. Contini, 'La questione del "Fiore" ', *CS* 13–14 (1965), 768–73; idem, 'Un'interpretazione di Dante', *Paragone* 188 (1965), 3–42 (also in *Un'idea di Dante* (n. 19 above), 69–111); idem, 'Un nodo della cultura medievale: La serie *Roman de la Rose–Fiore–Divina Commedia*', in *Un'idea di Dante*, 245–83; idem, *ED* s.v. *Fiore*; idem, ' "Mortuus ridivivus": Die Fiore-Frage', *DDJ* 61 (1986), 35–50; R. Fasani, *La lezione del 'Fiore'*, Milan, 1967 (for Folgore da San Gimignano, but see idem, *Il poeta del 'Fiore'*, Milan, 1971, and 'Il

"Fiore" e la poesia del Pucci', *DDJ* 49–50 (1974–5), 82–141, for Antonio Pucci); M. Muner, 'Perché il "Fiore" non può essere di Dante (e a chi invece potrebbe attribuirsi)' in *Motivi per la difesa della cultura* 7 (1968–9), 88–105, also ibid. 9 (1970–1), 274–320 (for Brunetto Latini); Aldo Vallone, *Dante*, Milan, 1971, 503–27 (for Alighieri); M. Picone, 'Il "Fiore": struttura profonda e problemi attributivi', *VR* 33 (1974), 145–56 (inclines to Alighieri); L. Vanossi, *Dante e il 'Roman de la Rose'. Saggio sul Fiore*, Florence, 1979 (fundamental; presupposes Dante's authorship).

Prior to Contini: Castets (the *Fiore*'s first editor, 1881), Mazzoni (Florence, 1901), D'Ovidio (1902 and 1907), Rajna (1921), Parodi (1922), Valli (1928), Ramacciotti (1936), and Köhler (1957) — all for Alighieri; D'Ancona (1881), Renier (1881), Gaston Paris (1895), Novati (1905), Farinelli (1908), Benedetto (1910), Zingarelli (1921), and Torraca (1921) — all against him.

22. L. J. Friedman, 'Jean de Meung, Antifeminism, and "Bourgeois Realism"', *MP* 57 (1959), 13–23; idem, 'Jean de Meun and Ethelred of Rievaulx', *L'Esprit Créateur* 2 (1962), 135–41; idem, 'Gradus Amoris', *RP* 19 (1965), 167–77; C. R. Dahlberg, 'Macrobius and the Unity of the *Roman de la Rose*', *STPH* 58 (1961) 573–82; idem, 'Love and the *Roman de la Rose*', *SP* 44 (1969), 568–84; J. V. Fleming, *The 'Roman de la Rose': A Study in Allegory and Iconography*, Princeton UP, 1969 — all along broadly 'Robertsonian' lines (see especially Robertson's *A Preface to Chaucer*, Princeton UP, 1962).

On the bourgeois, 'positive', and naturalistic interpretation of the *Rose* see, for example, E. Langlois, *Origines et sources du 'Roman de la Rose'*, Paris, 1890 (reprint Geneva, 1973), 98: 'Tel est l'esprit d'une partie de la littérature au treizième siècle, comprenant les fabliaux, le roman de Renart, une foule de poèmes de tous genres, qu'on pourrait grouper sous la dénomination commune de littérature satirique bourgeoise. A ce groupe appartient la seconde partie du *Roman de la Rose*'; L. Thuasne, *Le 'Roman de la Rose'*, Paris, 1929, 34: 'Jean de Meun est l'adversaire de l'amour courtois et platonique, non moins que des fadaises sentimentales qui en font la substance: il est positif et réduit l'amour aux plaisirs des sens'; and more recently G. Cohen, *Le 'Roman de la Rose'*, Besançon, 1973, 217: 'Nous avons insisté sur son naturisme robuste et jovial qui fait de lui l'adversaire le plus fougueux de l'hypocrisie et l'amour courtois et le premier et authentique représentant du naturalisme anticourtois, de l'esprit des fabliaux contre la poésie lyrique.'

23. L. F. Benedetto in *Il 'Roman de la Rose' e la letteratura italiana*, Beihefte zur *ZRP* 21, Halle, 1910, 106: 'Mi pare però che l'autore

italiano superi quello francese nello scetticismo, nell'incredulità schernitrice e beffarda. Entrambi hanno un contegno irriverente verso le cose sacre ed i religiosi'; G. Contini, 'Un nodo della cultura medievale' in *Un'idea di Dante* (n. 19 above), 272: 'Se questo mosaico ha un qualche significato, a me sembra che la sola soluzione possibile sia che il *Fiore* provenga dalla mano di Dante, ma, di più, che rappresenti qualche cosa di organico nella sua carriera, che costituisca, insomma, una prima prova di adattamento della *Rose*, un primo conato, conato tuttavia, diciamo pure, massimalistico, in quanto riduzione a un realismo esclusivo, *Roman* portato alla mera affabulazione . . . È l'introduzione di un ideale "borghese" che fa scomparire tutta la parte idealistica e cavalleresca'; E. Pasquini, *La letteratura didattica e la poesia popolare del Duecento*, Bari, 1971, 83: 'all'immagine idealizzata della civiltà cortese-cavalleresca (Guillaume scrive intorno alla terza decade del Duecento) s'oppone, quarant'anni più tardi, la condanna di quel mondo in nome della nuova ideologia laica e borghese'; L. Vanossi, *La teologia poetica del 'Detto d'Amore' dantesco*, Florence, 1974, 98: 'Ripercorrendo il cammino del secondo autore del *Roman de la Rose*, il poeta del *Fiore* tenta di conciliare la fedeltà ad Amore con la spregiudicata consapevolezza della 'realtà' dell'amore, da cui sono stati strappati i veli dell'illusione cortese.'

24. On Dante and 'translation', E. J. Richards, *Dante and the 'Roman de la Rose': An Investigation into the Vernacular Context of the 'Commedia'*, Tübingen, 1981.

   On Jean de Meun and Dante: the fourteenth-century Laurent de Premierfait (quoted by Fleming, n. 22 above, 18): '. . . peigny une vraye mapemond de toute choses celestes et terreines; Daut [i.e. Dante] donques, qui de Dieu et de Nature avoit receu l'esperit de poeterie, advisa que ou livre de la Rose est souffisammant descript le paradis des bons et l'enfer des mauvais en langaige francois, voult en langaige florentin, soubz aultre manier de verys rimoiez, contrefaire au vif le beau livre de la Rose, en ensuyant tel ordre comme fist le divin poete Virgile ou sixiesme livre que l'en nomme Eneide.'

25. On the date of the *Vita nuova*: M. Barbi, 'La data della *Vita nuova* e i primi germi della *Commedia*', in *Problemi di critica dantesca*, Florence, 1934, i. 99–112; M. Pazzaglia, *ED* s.v. *Vita nuova*.

   On the 'rifacimento' thesis—the idea that the final Beatrician chapters (XXXIX ff.) of the *Vita nuova* are a later addition dictated by the needs of the *Commedia*: L. Pietrobono, 'Il rifacimento della *Vita nuova* e le due fasi del pensiero dantesco', *GD* 42 (1941), 45–68 (with related essays in *Saggi danteschi*, Turin, 1954, 25–98), and B.

Nardi, 'Dalla prima alla seconda *Vita nuova*', in *Nel mondo di Dante*, Rome, 1944, 1–20; refuted by M. Barbi, 'Razionalismo e misticismo in Dante', in *Problemi di critica dantesca*, Florence, 1941, ii. 1–86, and M. Marti, 'Vita e morte della presunta doppia redazione della *Vita nuova*', in *Studi in onore di A. Schiaffini*, Rome, 1965, ii. 657–69. Also, U. Leo, 'Zum "rifacimento" der Vita Nuova', *RF* 84 (1967), 281–317.

26. Parallel passages: IV. 2: 'Ed io, accorgendomi del malvagio domandare che mi faceano, per la volontade d'Amore, lo quale mi comandava secondo lo consiglio de la ragione, rispondea loro che Amore era quelli che così m'avea governato'; XV. 8: 'ne la prima dico quello che Amore, consigliato da la ragione, mi dice quando le sono presso'; XXXIX. 2: 'lo mio cuore cominciò dolorosamente a pentere de lo desiderio a cui sì vilmente s'avea lasciato possedere alquanti die contra la costanzia de la ragione.' De Robertis (n. 16 above) 35–6: 'L'opposizione a Cavalcanti, con l'esplicita introduzione, proprio qui, di questo elemento da lui perentoriamente escluso ("Non è vertute, ma da quella vene Ch'è perfezione—ché se pone tale—*Non razionale, ma che sente*, dico. *For di salute* giudicar mantene, Ché la 'ntenzione per ragione vale": è la sua definizione d'amore, canz. *Donna me prega*, vv. 29–33), e proprio nel momento che ne è riconosciuta la straordinaria presenza poetica, non potrebbe essere più chiara. E questo è, sul piano letterario come su quello ideologico, il motivo informatore del libretto.'

27. Augustine, *Conf.* I. vii; III. xii; IX. viii, for the author as editor. Ibid. x. viii, for the rubrics in memory motif. Jean de Meun, *Roman de la rose*, 15115–20, for the author as glossator. The 'inner meaning' or *sententia* motif may owe something to the 'id quod libris pretium facit' passage in Boethius, *Consol.* I. pr. v.

28. See, for example, Richard of St Victor, *Ben. maj.* v. v (*PL* 196, col. 174): 'Magnitudine jocunditatis, et exsultationis mens hominis a seipsa alienatur, quando intima illa internae suavitatis abundantia potata, imo plene inebriata, quid sit, quid fuerit, penitus obliviscitur, et in abalienationis excessum, tripudii sui nimietate traducitur, et in supermundanum quemdam affectum, sub quodam mirae felicitatis statu raptim transformatur.' There is a useful translation of Richard of St Victor (including the *Benjamin minor* and *major*) in the SPCK Classics of Western Spirituality series, London, 1979 (preface by J. Châtillon).

29. 'The poet may even know that this figure of the circle is one long used to designate the Deity according to His undeniable attribute of transcendental omniscience. And Love is a God and as such may be said to be as the centre of a circle in that he is able to see all the

points on the line of time, past, present, and future, as if they were points on the circumference of a circle and hence all equidistant from him. Which would mean that the God of Love is saying that he is a God and that the poet is not *(tu autem non sic)*' (Singleton, *An Essay on the Vita Nuova*, The Johns Hopkins UP, 1977 (1949), 17); 'Ma è pur verosimile che l'immagine del cerchio adombrasse il concetto della perfezione (come indifferenziata coerenza, come assoluta similitudine) dell'idea rispetto alla approssimatività e dissimilitudine e, per così dire, dissimmetria dell'operare umano, moventesi non in ordine al suo principio, ma secondo direzioni avventurose ed occasionali. E non escluderei che il rapporto ivi formulato (". . . cui simili modo se habent circumferentie partes") racchiudesse un invito a ristabilire quella intima "proportio", a riconformare ciascuna parte al tutto' (De Robertis, *Il libro della Vita Nuova*, 2nd edn., Florence, 1970, 69).

Boethius, *Consol.* IV. pr. vi; Richard of St Victor, *Ben. maj.* II. xx (*PL* 196, col. 101); Aquinas, *ScG* I. lxvi; Bonaventure, *It. men.* v. 8. Important tonally, however, the 'Vidit enim se stantem in quadam regula lignea' passage of Augustine, *Conf.* III. xi (Monica's dream).

30. Cf. *Roman de la rose* 1871–8; 4263 ff.; Augustine, *Conf.* VII. v.
31. On the *gabbo* (*gab*): G. Barberi Squarotti, 'L'interpretazione del "gabbo"' in *L'artificio dell'eternità*, Verona, 1972, 107–29; M. Picone, '*Vita Nuova*' *e tradizione romanza*, Padua, 1979, 99–128 (with bibliography).
32. Cf., for the 'nuova matera' motif, *Roman de la rose*, 39; 2058–64.
33. Cf. Aquinas *In eth.* (n. 1576); *ScG* I. xci. For Cicero—in addition to *De amic.* ix. 31 passage cited in the text—ibid. viii. 27; xxi. 79–80; xxvii. 100; *De fin.* II. xxiv. 78, etc.; Seneca, *Ep. mor.* VI. 2; Jean de Meun, *Roman de la rose*, 4739–52; Brunetto Latini, *Tres.* II. 43 and 102 ff.

G. Vansteenberghe, 'Deux théoriciens de l'amitié au XIIe siècle: Pierre de Blois et Aelred de Rieval', *Revue des sciences religieuses* 12 (1932), 572–88; J. Leclercq, 'L'amitié dans les lettres au moyen âge', *RMAL* I (1945), 391–410; P. Delhave, 'Deux adaptations du De amicitia de Cicéron au XIIe siècle', *Recherches de théologie ancienne et médiévale* 15 (1948), 304–31; A. M. Fiske, *Friends and Friendship in the Monastic Tradition*, Mexico, 1970. Translation of Aelred of Rievaulx, *Spiritual Friendship*, M. E. Laker, Cistercian Fathers Series No. 5, Washington DC, 1974.

On Dante and friendship: C. Singleton, *An Essay on the Vita Nuova*, The Johns Hopkins UP, 1977 (1949), *passim*; D. De Robertis (n. 16 above), 107 ff.

On spiritual friendship in Jean de Meun—he claims to have

translated Aelred (*Li Livres de Confort de Philosophie*, ed. V. L. Dedeck-Héry, *MS* 14 (1952), 165–275 at 168: '. . . et translatay de latin en francois . . . le livre Aered de Esperituelle Amitié')—see L. J. Friedman, 'Jean de Meun and Ethelred of Rievaulx', *L'Esprit Créateur* 2 (1962), 135–41.

34. The notion is Ciceronian (though cf. 1 Tim. 6: 17: 'neque sperare in incerto divitiarum'): *De amic.* ii. 7; vi. 20; viii. 30; *De fin.* II. xiv. 46; Seneca, *De benef.* I. v; *Ep. mor.* VIII. 3 ff.; IX. 16 ff.; XXVII. 3.

35. 'Porro unum est necessarium. Maria optimam partem elegit quae non auferetur ab ea'; Seneca, *Ep. mor.* IX. 19; Augustine, *De vera rel.* xlvi. 86; Richard of St Victor, *Ben. maj.* I. i (*PL* 196, cols. 64–5); Brunetto Latini, *Tres.* II. 104. 1.

36. 'Indeed, now we may see that this is actually no resemblance of the person of Beatrice to that of Christ. Neither in character nor in appearance would the one suggest the other. Nor would we expect this to be so. Between the two is an infinite gulf, the distance between God and the creature. No principle of close resemblance of any creature to Christ will hold in view of that distance. This resemblance results rather from a likeness to be seen not between two persons but between two actions: that is, the action in which Beatrice has the role which her name itself implies (a bringer of beatitude) is like the action in which Christ has such a role. Both are actions leading to *salute*, to the beatitude of Heaven. And in the one Beatrice *is*, as Christ *is* in the other: a love which comes from Heaven and returns, *through* death, to Heaven. The actions are two, the roles in the actions are two, and no degree of equivalence is affirmed' (C. S. Singleton, n. 29 above, 112). Firmer, however, in his sense of the non-specifically Christian nature of Dante's solution in the *Vita nuova*, De Robertis (n. 16 above), 121–2: 'Ma la *Vita Nuova* testo di esperienza mistica non è; anzi, tutt'al contrario di quello che dice il Singleton, evidentemente sotto l'influenza della contrapposizione istituita dal Gilson nel suo scritto su *Saint Bernard et l'amour courtois*, essa si muove ancora nell'ordine della concezione cortese. La donna è ancora l'oggetto ultimo dell'amore. Di mutato, rispetto alla concezione cortese, non è l'oggetto, ma l'amore, che ha assunto i caratteri della "Dei dilectio". È l'esperienza mondana, cioè, che è ricreata dall'esperienza cristiana; ma essa non si struttura per gradi, in forma d'ascesi, bensì come storia, interamente nella sfera mondana.' On the *Vita nuova* as an uncomfortable amalgam of the Christian and 'courtly', N. Mineo, *Dante*, Letteratura Italiana Laterza, Bari, 1971, 55–6: 'Non vogliamo dire tuttavia che la *Vita Nuova* sia un'opera assolutamente perfetta. Il suo limite consiste

però in qualcosa di interno alla sua struttura e alla poetica di Dante in quel tempo, in un certo che di ambiguo derivante dalla contaminazione di tematica mondano-cortese e di tematica religiosa.'
On analogy: H. Lyttkens, *The Analogy between God and the World*, Uppsala, 1952; R. M. McInerny, *The Logic of Analogy: an Interpretation of St Thomas*, The Hague, 1961.

37. Cf. the 'expetebant eum undique oculi mei' passage in Augustine, *Conf.* IV. iv. On the 'concupiscence of the eyes', Richard of St Victor, *Ben. maj.* III. iii (*PL* 196, col. 113).

CHAPTER 2

1. In general: C. T. Davis, 'Education in Dante's Florence', *SP* 40 (1965), 415–35, now in *Dante's Italy and Other Essays*, Philadelphia UP, 1984, 137–65; S. Vanni Rovighi, 'Le "disputazioni de li filosofanti"', in *Dante e Bologna*, Bologna, 1967, 179–92; H. Wieruszowski, 'Rhetoric and the Classics in Italian Education of the Thirteenth Century', *Studia Gratiana* 11 (1967), 169–208, and in *Politics and Culture in Medieval Spain and Italy*, Rome, 1971, 589–627.

On Santa Maria Novella (Dominican): I. Taurisano, 'L'organizzazione delle scuole domenicane nel secolo xiii', in *Miscellanea lucchese di studi storici e letterari in memoria di Salvatore Bongi*, Lucca, 1931, 93–130; S. Orlandi, *La biblioteca di S. Maria Novella in Firenze dal sec. xiv al sec. xix*, Florence, 1952; G. Pomaro, 'Censimento dei manoscritti della biblioteca di S. Maria Novella. Parte I: Origini e Trecento', *Memorie domenicane* NS 11 (1980), 325–470. For Remigio de' Girolami, M. Grabmann, 'La scuola tomistica italiana del XIII e del principio del XIV secolo', *RFNS* 15 (1923), 97–155 (also in his *Mittelalterliches Geistesleben*, Munich, 1926, i. 332–91); idem, 'Fra Remigio de' Girolami O. Pr., discepolo di San Tommaso d'Aquino e maestro di Dante', *La scuola italiana* 53 (1925), 267–81, 347–68; L. Minio-Paluello, 'Remigio de' Girolami's *De Bono Communi*: Florence at the Time of Dante's Banishment', *ITST* 11 (1956), 56–71; C. T. Davis, 'Remigio de' Girolami and Dante: A Comparison of their Concepts of Peace', *SD* 36 (1959), 105–36; idem, 'An Early Florentine Political Theorist: Fra Remigio de' Girolami' (1960), now in *Dante's Italy* (above), 198–223 (with a useful update on Remigio studies generally, 320–1); A. Sarubbi, *Chiesa e stato comunale nel pensiero di Remigio de' Girolami*, Naples, 1971; C. D. Matteis, *La teologia politica comunale di Remigio de' Girolami*, Bologna, 1977.

On Santa Croce (Franciscan): F. Mattesini, 'La biblioteca di S. Croce e Fra Tedaldo della Casa', *SFRAN* 57 (1960), 254-316; C. T. Davis, 'The Early Collection of Books of S. Croce in Florence', *Proceedings of the American Philosophical Society* 107 (1963), 399-414. For Pier di Giovanni Olivi and Ubertino da Casale (lectors in Santa Croce between 1287 and 1289), R. Manselli, 'Pietro di Giovanno Olivi ed Ubertino da Casale'. *SM* 3rd series) 6 (1965), 95-122; idem, 'Firenze nel Trecento: Santa Croce e la cultura francescana', *Clio* 9 (1973), 325-42; and more generally M. Reeves, *The Influence of Prophecy in the Later Middle Ages*, Oxford, 1969; idem, *Joachim of Fiore and the Prophetic Future*, London, 1976.

On Santo Spirito (Augustinian): D. Gutierrez, 'La biblioteca di Santo Spirito in Firenze', *Analecta Augustiniana* 25 (1962), 5-88.

On Dante and Brunetto Latini (as teacher): H. Wieruszowski, 'Brunetto Latini als Lehrer Dantes und der Florentiner', *Archivio italiano per la storia della pietà* 2 (1958-9), 169-98, and in *Politics and Culture* (above) 515-61; C. T. Davis, 'Brunetto Latini and Dante', *SM* (3rd series) 8 (1967), 421-50, and in *Dante's Italy* (above) 166-97; F. Mazzoni (in addition to his entry in *ED* s.v. (iii. 579-88), 'Brunetto in Dante' (preface to the *Tesoretto* and *Favolello*), Alpignano, 1967. Also, with reference to Brunetto Latini's place in hell, P. Armour, 'Dante's Brunetto: the Paternal Paterine', *ITST* 38 (1983), 1-38.

2. Foster-Boyde, *Dante's Lyric Poetry*, OUP, 1967, ii. 161.
3. G. Busnelli, 'Un famoso dubbio di Dante intorno alla materia prima', *SD* 13 (1928), 47-60.
4. M. Bowra, 'Dante and Arnaut Daniel', *SP* 27 (1952), 459-74; L. Blasucci, 'L'esperienza delle "petrose" e il linguaggio della "Divina Commedia"', *BEL* 12 (1957), 403-31; E. Melli, 'Dante e Arnaut Daniel', *FR* 6 (1959), 423-48; M. Marti, 'Sulla genesi del realismo dantesco', in *Realismo dantesco e altri studi*, Milan-Naples, 1961, 1-32; idem, 'Le rime realistiche (La tenzone e le petrose dantesche)', *NLD* 8 (1976), 209-30; S. Battaglia, *Le rime 'petrose' e la sestina (Arnaldo Daniello—Dante—Petrarca)*, Naples, 1964; E. Fenzi, 'Le rime per la donna Pietra', in *Miscellanea di studi danteschi*, Geneva, 1966, 229-309; I. Baldelli, 'Ritmo e lingua di "Io son venuto al punto de la rota"', in *Critica e storia letteraria. Studi offerti a M. Fubini*, Padua, 1970, i. 342-51; P. Bondanella, 'Arnaut Daniel and Dante's Rime Petrose: A Re-examination', *STPH* 68 (1971), 416-34; M. Perugi, 'Arnaut Daniel in Dante', *SD* 51 (1974-5), 74-80; B. Comens, 'Stages of Love, Steps to Hell: Dante's Rime Petrose', *MLN* 1.1 (January 1986), 157-88. More generally, S. Santangelo, *Dante e i trovatori provenzali*, Catania, 1921 (2nd edn., 1959).

On the Forese *tenzone*: M. Barbi, *Problemi di critica dantesca*, Florence, 1941, ii. 87–214; F. Chiappelli, 'Proposta d'interpretazione per la tenzone di Dante con Forese Donati', *GSLI* 142 (1965), 321–50.

5. J. Boutière and A. Schutz, *Biographies des troubadours*, Paris, 1964, 59: 'e pres una maniera de trobar en caras rimas, per que soas cansons no son leus ad entendre ni ad aprendre.'
The text and translation of *Sols sui* are from the *Anthology of Troubadour Lyric Poetry*, ed. A. R. Press, Edinburgh UP, 1971, 186–8.

6. Contini, *Poeti del Duecento*, Milan–Naples, 1960, ii. 548–9. M. Santagata, 'Lettura cavalcantiana (XLI "I' vegno 'l giorno a te 'nfinite volte")', *GSLI* 148 (1971), 295–308; M. Ciccuto, 'I sonetti di Guido Cavalcanti a Dante', *RANL*, ser. viii, 32 (1977), 399–434.

7. A. Jeanroy, 'La "sestina doppia" de Dante et les origines de la sextine', *Romania* 42 (1913), 481–9.

8. Andreas Capellanus, *De amore* I. vi (Battaglia, 180; Walsh, 156); II. iv (Battaglia, 288; Walsh, 232); II. viii (Regula III); *Roman de la rose*, 2227–32; *Detto d'amore*, 449–54. 'Il *conflictus*', says Contini in his edition of the *Rime*, 111, 'è composto nella convivenza di piacere e d'azione, a cui quelle [i.e. Bellezza and Virtù] rispettivamente conducono; e così questo sonetto è indicativo dell'ideale supremo che Dante si propone verso i trent'anni, la fusione dell'eleganza mondana e della *rectitudo*, delle qualità laiche e chiericali insieme: quella che gli consentiva di far l'elogio della virtù assoluta e della virtù relativa del cavaliere.' On *Poscia ch'Amor*, F. Montanari, *La canzone della leggiadria: 'Poscia ch'Amor del tutto m'ha lasciato'*, Turin, 1961.

9. On the possibility of these poems' post-dating Dante's exile, C. Grayson, *MAEV* 37 (1968), 317–18.

10. On *Tre donne*: M. Barbi, 'Per l'interpretazione della canzone "Tre donne"', *Problemi di critica dantesca*, Florence, 1941, ii. 267–76; M. Casella, 'Tre donne intorno al cor mi son venute', *SD* 30 (1951), 6–22; K. Foster, 'Dante's canzone "Tre donne"', *ITST* 9 (1954), 56–68, also in *God's Tree*, London, 1957, 15–32.
On *Doglia mi reca*: P. Boyde, 'Style and Structure in Dante's canzone, *Doglia mi reca*', *ITST* 20 (1965), 26–41 (also, modified, in *Dante's Style in his Lyric Poetry*, CUP, 1971, 317–31); V. Pernicone, *ED* s.v. (ii. 531–2); M. Picone, 'Giraut de Bornelh nella prospettiva di Dante', *VR* 39 (1980), 22–43.

11. Cf., iconographically, Tritesce in the *Roman de la rose*, 313–24. Similarly dejected in tone is the political sonnet *Se vedi li occhi miei*, dating from the same period (the second quatrain relates probably to Clement V and to Philip the Fair of France):

> Se vedi li occhi miei di pianger vaghi
> per novella pietà che 'l cor mi strugge,
> per lei ti priego che da te non fugge,
> Signor, che tu di tal piacere i svaghi:
>
> con la tua dritta man, cioè, che paghi
> chi la giustizia uccide e poi rifugge
> al gran tiranno, del cui tosco sugge
> ch'elli ha già sparto e vuol che 'l mondo allaghi;
>
> e messo ha di paura tanto gelo
> nel cor de' tuo' fedei che ciascun tace.
> Ma tu, foco d'amor, lume del cielo,
>
> questa vertù che nuda e fredda giace,
> levala su vestita del tuo velo,
> ché sanza lei non è in terra pace.

12. A. Pézard, '"De passione in passionem": Cino, *Dante, quando per caso s'abbandona* (*Rime* cx–cxi)', *Alighieri* 1 (1960), 14–26; also idem, 'Le sonnet de la Dame Verte (Dante a Cino, *Rime* xcv)', *Revue des études italiennes*, NS 11 (1965), 329–80.

    On Cino as a political theorist, and on the shifts in his political theory, see ch. 4 n. 13 below (Maffei especially).

13. Andreas Capellanus, *De amore* II. 5 (Battaglia, 304; Walsh, 244).

14. F. Maggini, 'La canzone montanina', in *Studi letterari. Miscellanea in onore di E. Santini*, Palermo, 1956, 95–100; C. G. Hardie, 'Dante's "Canzone Montanina"', *MLR* 55 (1960), 359–70; F. Montanari, 'La canzone "Amor, da che convien pur ch'io mi doglia"', *LM* 12 (1962), 359–68.

15. Aquinas, *ScG* I. iv; Cicero, *De off.* I. iv. 13; Richard of St Victor, *Ben. maj.* I. xii, ult. (*PL* 196, coll. 78–80). M. P. Simonelli, 'Pubblico e società nel *Convivio*', *Yearbook of Italian Studies* 4 (1980), 41–58.

16. I. ii. 1–2: 'Per che io, che ne la presente scrittura tengo luogo di quelli, da due macule mondare intendo primieramente questa esposizione, che per pane si conta nel mio corredo. L'una è che parlare alcuno di se medesimo pare non licito'; Prov. 27: 2; Cicero, *De off.* I. xxxviii. 137; Capellanus, *De amore* I. vi (Battaglia, 26; Walsh, 48); Aquinas, *ST* IIa IIae, 109, 1 ad 2. Augustine, by contrast, on the social value of confession, *Conf.* x. iv.

17. The notion is Sapiential (Wisdom 6 ff., etc.) but see too Augustine, *Conf.* III. iv. The assimilation motif looks Boethian (*Consol.* III. pr. x). The etymology of the term 'philosophy' is commonplace.

18. In fact, Aristotle, *Pol.* IV. 8, 1294ª21 and Aquinas ad loc. Cf. *Mon.* II. iii. 4.
19. Jean de Meun, *Roman de la rose*, 4945–60; 5041–60; 17531–8; Boethius, *Consol.* II. met. v. 27–30; III. pr. iii; Cicero, *De fin.* II. xxvii. 86; Seneca, *Ep. mor.* XC. 41; Augustine, *Conf.* IV. vi; Bernard of Clairvaux, *De dil. Deo* vii (*PL* 182, col. 985C); Aquinas, *ScG* III. xxx; cxxxii. 7, etc.

    On lineage (*Conv.* IV. vii. 8–9: 'Quale di costoro si dee dicere valente? Rispondo: quelli che andò dinanzi. Questo altro come si chiamerà? Rispondo: vilissimo. Perché non si chiama non valente, cioè vile? Rispondo: perché non valente, cioè vile, sarebbe da chiamare colui che, non avendo alcuna scorta, non fosse ben camminato; ma però che questi l'ebbe, lo suo errore e lo suo difetto non può salire, e però è da dire non vile, ma vilissimo. E così quelli che dal padre o d'alcuno suo maggiore buono è disceso ed è malvagio, non solamente è vile, ma vilissimo, e degno d'ogni dispetto e vituperio più che altro villano'); Jean de Meun, *Roman de la rose*, 18577–604; 18725–31.
20. On health in body and mind: *De off.* I. xxxiv. 122; on attention to seniority: *De senec.* viii. 26; *De off.* I. xxxiv. 122 (but also Prov. 1: 8; Col. 3: 20, etc.); on wise counsel: *De off..* I. xxxiv. 123; *De senec.* ix. 29 (but also Aristotle, *Nic. eth.* VI. 11 (1143ᵇ11–13), and Aquinas ad loc. (n. 1254), etc.).
21. Also Ciceronian (*De senec.* xix. 71).
22. Useful, accessible, and authoritative short histories of medieval philosophy include: G. Leff, *Medieval Thought: St Augustine to Ockham*, London (Penguin), 1958; F. C. Copleston, *A History of Philosophy*, vol. 2 (Mediaeval Philosophy), part II (Albert the Great to Duns Scotus), New York, 1962 (1950); idem, *A History of Medieval Philosophy*, London, 1972; D. Knowles, *The Evolution of Medieval Thought*, London, 1962; A. A. Maurer, *Medieval Philosophy*, New York, 1962.

    On Dante in particular: M. Barbi, 'Razionalismo e misticismo in Dante' in *Problemi di critica dantesca* (n. 4 above), ii. 1–86; B. Nardi, *Dante e la cultura medievale*, Bari, 1942; idem, *Nel mondo di Dante*, Rome, 1944; idem, *Dal Convivio alla Commedia*, Rome, 1960; idem, *Saggi e note di critica dantesca*, Milan–Naples, 1966 (with an important general survey, 3–109); idem, *Saggi di filosofia dantesca*, 2nd edn., Florence, 1967 (1930)—all fundamental; E. Gilson, *Dante and Philosophy*, New York, 1963 (Paris, 1939); E. Garin, 'Il pensiero di Dante', in *Storia della filosofia italiana*, 2nd edn., Turin, 1966, i. 179–206; K. Foster, *The Two Dantes and Other Studies*, London, 1977 (especially 156–253); L. Minio-

Paluello, 'Dante's Reading of Aristotle', in *The World of Dante*, ed. C. Grayson, OUP, 1980, 61–80; M. Corti, *Dante a un nuovo crocevia*, Florence, 1982 (especially 77–101); idem, 'La filosofia aristotelica e Dante', *LC* 13 (1984), 109–23.

23. On early Latin Averroism (in addition to the general histories, n. 22 above): P. Mandonnet, OP, *Siger de Brabant et l'Averroisme latin au XIIIe siècle*, Louvain, 1911; M. Grabmann, 'L'aristotelismo italiano al tempo di Dante con particolare riguardo all'Università di Bologna', *RFNS* 38 (1946), 260–77; B. Nardi, 'L'averroismo bolognese nel secolo XIII e Taddeo Alderotto', *RSF* 4 (1949), 11–22; M. Corti, *Dante a un nuovo crocevia* (n. 22 above). Boethius of Dacia's *De summo bono* (interesting as a term of comparison for the *Convivio*) is in the *Corpus Philosophorum Danicorum Medii Aevi*, vi (1976), 1–2 (Opuscula), ed. N. J. Green-Pedersen and J. Pinborg (and in M. Grabmann, *Mittelalterliches Geistesleben*, Munich, 1936, ii. 200–24).

Useful (again in addition to the histories) on Aquinas: A. C. Pegis, *Introduction to St Thomas Aquinas*, New York, 1948 (readings); E. Gilson, *Le Thomisme*, Paris, 1948; idem, *The Elements of Christian Philosophy*, New York, 1960; F. Copleston, *Aquinas*, London, 1955 (Penguin).

24. 'La dottrina dell'Empireo', in *Saggi di filosofia dantesca* (n. 22 above), 214.

25. For Dante's astronomy and cosmology: M. A. Orr, *Dante and the Early Astronomers*, rev. edn., London, 1956 (1913); E. Moore, 'The Astronomy of Dante', in *Studies in Dante*, ser. iii, London, 1968 (1903), 1–108 (109–43 are on Dante's geography); P. Boyde, *Dante Philomythes and Philosopher: Man in the Cosmos*, CUP, 1981.

26. Albert the Great, *De coelo* II. 3. 11; *Summa de creaturis* III. 12. 3; *ST* II. 11. 52. 3. B. Nardi, 'Dante e Alpetragio' in *Saggi di filosofia dantesca* (n. 22 above), 153 ff.

27. *Liber de causis*, propp. 3 a and, especially, 8 a; Aquinas ad loc. (in such a way, however, as to reserve creativity for God alone). S. Bemrose, *Dante's Angelic Intelligences*, Rome, 1983.

28. Cf. Cicero, *De amic.* iv. 14; *De senec.* xxi. 77; Seneca, *Ep. mor.* LXV. 16; Boethius, *Consol.* III. met. xi. 9–10; Aquinas (against Plato, the Manichees, and Origen), *ScG* II. lvii; lxxxiii; etc.

29. Aristotle, *Nic. eth.* I. 6 (1098^a19); Aquinas (who extends the 'vita perfetta' motif into eternity) ad loc. nn. 129–30; *ScG* III. lxiii; Cicero, *De fin.* II. vi. 19.

30. E. Gilson, *Dante and Philosophy* (n. 22 above), 99 ff. ('The Primacy of Ethics').

31. K. Foster, 'The Two Dantes' (n. 22 above), 159: 'Thus we see

Dante in the *Convivio* coming out with ideas about the perfectibility of man in this life, and of the soul in the next, without its apparently crossing his mind that he was begging the question (from the standpoint of orthodox Christianity) as to whether man *could* reach perfection, here or hereafter, unassisted by divine grace. In the *Convivio* the Christian doctrine of grace—and so of man's *de facto* inherent sinfulness and natural incapacity to bring himself, by his own effort, to union with God—is virtually ignored.' Recently on prevenient grace in the *Commedia*, P. Boyde, 'Predisposition and Prevenience: Prolegomena to the Study of Dante's Mind and Art', *Proc. British Academy* 69 (1984) (Italian Lecture, 1983).

32. *ScG* III. xliii (for Aquinas's account of Averroistic intellectual happiness here and now).

33. A neat instance of the way in which Dante received and adapted for his own purposes arguments central to his *auctores*, in this case Aquinas, *ScG* III. xlviii.

34. e.g. I. 9 (1099^b25); IX. 9 (1169^b28), etc., and Aquinas ad loc. (n. 1895, etc.). Suggestive too is Giles of Rome, *De reg. prin.*, *passim* (e.g. I. i. 4: 'posuerunt enim felicitatem politicam et contemplativam: ut dicatur quis felix politice quando est felix ut homo, habendo in se prudentiam, quae est recta ratio agibilium; dicatur felix contemplative, quando est supra hominem, et quando est felix non solum ut homo, sed ut est in eo aliquid divinum et aliquid melius homine . . .'—where we catch a glimpse of one or two other Dantean motifs: the more-than-human motif and the at-one-with-the-angels motif).

35. Further on Dante's providential re-reading of Virgil in the *Convivio*: U. Leo, 'The unfinished *Convivio* and Dante's re-reading of the *Aeneid*', in *Sehen und Wirklichkeit bei Dante*, Frankfurt am Main, 1957, 71–104; D. Thompson, 'Dante and Bernard Silvestris', *Viator* I (1970), 201–6.

36. e.g. *De. civ. Dei* IV. iii–iv—the famous 'a state without justice is no state at all' chapters.

37. Augustine, *Enchir.* 48, 51, etc.; *De peccat. mer. et rem.* 39 ff. (and the anti-Pelagian writings generally). Aquinas, by contrast, on sin as *destitutio* or deprivation, as distinct from vitiation, *ST* Ia IIae, 85, I conc. B. Nardi, 'Il concetto dell'Impero' in *Saggi di filosofia dantesca* (n. 22 above), 215–28.

38. Busnelli and Vandelli (ed. *Conv.* II. 99, n. 6) point to *De lib. arb.* I. xxxi and *Ennar. in Ps.* I n. 2, but note (with Nardi) that Dante may be merely paraphrasing or recalling from memory.

39. On the prophetic aspects of the *Commedia*: B. Nardi, 'Dante

profeta', in *Dante e la cultura medievale* (n. 22 above), 258–334; N. Lenkeith, 'The Poet as Prophet', in *Dante and the Legend of Rome*, The Warburg Institute, London, 1952, 32–72; N. Mineo, *Profetismo e Apocalittica in Dante*, Catania, 1968; R. Morghen, 'Dante profeta', *LC* 3 (1970), 13–36; M. Reeves, 'Dante and the Prophetic View of History', in *The World of Dante: Essays on Dante and His Time*, OUP, 1980, 44–60; G. Gorni, 'Spirito profetico duecentesco e Dante', *LC* 13 (1984), 49–68; A. Piromalli, 'Messaggi politici, simboli, profezie nella *Commedia*, *LC* 16 (1987), 29–50.

40. A. Vallone, *La prosa del 'Convivio'*, Florence, 1967; C. Segre, *Lingua stile e società*, Milan, 1976 (1963), 227–70.
41. On beneficence: Aristotle, *Nic. eth.* IV. 1 ($1119^{b}20$–$22^{a}18$) and St Thomas ad loc. (nn. 649–705); Seneca (cited, though possibly at second hand, at I. viii. 16), *De ben.* I. vi. 1; Cicero, *De amic.* i. 4; *De fin.* III. xx. 65; Augustine, *De doct. christ.* III. x. 16; *Conf.* XI. ii; Brunetto Latini (for whom liberality coincides with *cortoisie* and is a branch of justice), *Tres.* II. 94 and 97.
42. For sunsets and sunrises, Richard of St Victor, *Ben. maj.* III. vi (*PL* 196, col. 117), after Esther 8: 16; Is. 9: 2; Luke 1: 79, etc.; for breadbaskets, Augustine, *De doct. christ.* I. i. 1, (after Matt. 14: 17–21; 15: 34–8).

CHAPTER 3

1. Cf. *mutatis mutandis* (Latin with respect to Greek), Cicero, *De fin.* I. iii. 10; III. ii. 5; and especially III. xii. 40.
2. Augustine, *De doct. christ.* II. ii. 3; *De trin.* X. 1. 2.
3. Augustine, *De doct. christ.* I. vi. 6. D. Castaldo, 'L'etica del primiloquium di Adamo nel *De vulgari eloquentia*', *IT* 59 (1982), 3–15.
4. Also *De doct. christ.* prooem. 5.
5. Augustine, *De doct. christ.* II. iv. 5. L. Sebastio, 'De vulgari eloquentia I. vii. 4–7', *Alighieri* 23 (1982), 18–44.
6. Ut silvae foliis pronos mutantur in annos,
   prima cadunt, ita verborum vetus interit aetas,
   et iuvenum ritu florent modo nata vigentque.
   Debemur morti nos nostraque . . .

   Multa renascentur quae iam cecidere, cadentque
   quae nunc sunt in honore vocabula, si volet usus,
   quem penes arbitrium est et ius et norma loquendi.

   Seneca, *Ep. mor.* CXIV. 13; Augustine on mutability, *De doct. christ.* I. viii. 8 ult.

Parallel passage for *Conv.* I. v. 9–10 and *DVE* I. ix. (though with a greater sense of the *naturally*, as distinct from *unnaturally*, evolutionary character of language): *Par.* XXVI. 124–38.

7. Cf. Giles of Rome, *De reg. prin.* II. ii. 7: 'Videntes enim Philosophi nullum idioma vulgare esse completum et perfectum, per quod perfecte exprimere possent naturam rerum, et mores hominum, et cursus astrorum, et alia de quibus disputare volebant, invenerunt sibi quasi proprium idioma, quod dicitur latinum, vel idioma literale: quod constituerunt adeo latum et copiosum, ut per ipsum possent omnes suos conceptus sufficienter exprimere.'

   On Italian in relation to Latin: C. Grayson, 'Nobilior est vulgaris', in *Centenary Essays on Dante*, OUP, 1965 (with reference to Marigo and Vinay). On speculative grammar in the Middle Ages: M. Corti, *Dante a un nuovo crocevia*, Florence, 1982, 9–76, and G. C. Alessio, 'La grammatica speculativa e Dante', *LC* 13 (1984), 69–88.

8. Aquinas, *ScG* I. xxviii (for the form of the argument).
9. Boethius, *Consol.* IV. pr. vi (for the imagery).
10. See in general *The New Grove Dictionary of Music and Musicians*, London, 1980, s.v. 'Troubadours', 'trouvères' (xix. 201–3). Also P. Aubry, *La Rythmique musicale des troubadours et des trouvères*, Paris, 1906; J. Beck, *Die Melodien der Troubadours*, Strasbourg, 1908; R. Monterosso, *Musica e ritmica dei trovatori*, Milan, 1956.
11. Similarly, Cicero, *De orat.* III. xlv. 177; lii. 199; lv. 212; Quintilian, *Inst. orat.* XII. x. 58; Geoffrey of Vinsauf, *Doc.* (ed. Faral), 145; John of Garland, *Paris. poet.* I. 124–34 and, especially, II. 116–23—the *rota Virgilii* passage (ed. Lawler, Yale UP, 1974, 10 and 38 ff.). E. Norden, *Die antike Kunstprosa*, Stuttgart, 1958 (1898); F. Quadlbauer, *Die antike Theorie der 'genera dicendi' im lateinischen Mittelalter*, Vienna, 1962.
12. John of Garland, *Paris. poet.* IV. 477–81 (Lawler, 80–2); Uguccione, *Deriv.* ad loc.; Dante, *Ep.* XIII. 10.
13. On the *cursus*, in addition to Toynbee, ed. *Epistolae*, Oxford, 1966 (1920), 224–47 (Appendix C), and Mengaldo, *ED* s.v.: A. C. Clark, *The Cursus in Medieval and Vulgar Latin*, Oxford, 1910; M. G. Nicolau, *L'origine du "cursus" rythmique et les débuts de l'accent d'intensité en latin*, Paris, 1930; A. Marigo, 'Il "cursus" nel *De vulgari eloquentia* di Dante', *Atti e memorie dell'Accademia di scienze, lettere, ed arti di Padova*, NS 48 (1931–2), 85–112; P. Rajna, 'Per il "cursus" medievale e per Dante', *SFI* 3 (1932), 7–86; F. di Capua, 'Appunti sul "cursus" o ritmo prosaico nelle opere latine di Dante Alighieri', and 'Lo stile isidoriano nella retorica medievale e in Dante', in *Scritti minori*, Rome, 1959, i. 564–85 and ii. 226–51.

Also, more generally, G. Lindholm, *Studien zum mittellateinischen Prosarhythmus*, Stockholm, 1963.

On the *ars dictaminis*, C. H. Haskins, 'The Early "Artes dictandi" in Italy', in *Studies in Mediaeval Culture*, Oxford, 1929, 170–92; H. Wieruszowski, '"Ars dictaminis" in the time of Dante', *MH* 1 (1943), 95–108 (now in *Politics and Culture in Medieval Spain and Italy*, Rome, 1971, 359–77); F. J. Schmale, 'Die Bologneser Schule der Ars Dictandi', *Deutsche Archiv für Erforschung des Mittelalters namens der MGH* 13 (1957), fasc. i; G. Vecchi, *Il magistero delle "Artes" latine a Bologna nel medioevo*, Bologna, 1958. On Brunetto Latini and the *ars dictaminis*, G. C. Alessio, 'Brunetto Latini e Cicerone (e i dettatori)', *IMU* 22 (1979), 123–69. Texts of, among others, Alberic of Monte Cassino, Hugh of Bologna, Buoncompagno, and Guido Faba are contained in L. Rockinger (ed.), *Briefsteller und Formelbücher des eilften bis vierzehnten Jahrhunderts*, 2 vols., Munich, 1863. The *Ars dictaminis* of Giovanni del Virgilio is edited by P. O. Kristeller in *IMU* 4 (1961), 181–200, and Guido Faba's *Summa dictaminis* by A. Gaudenzi, *Il Propugnatore*, NS 3 (1890), 287–338 and 345–93. Early Florentine formularies are in *Formularium Florentinum Artis Notariae (1220–1242)*, ed. G. Masi, Milan, 1943.

14. *Ad Herenn.* IV. xxxiv. 45; Isidore, *Etym.* I. xxxvii. 2; Geoffrey of Vinsauf, *Poetria nova*, 774. For the other figures mentioned here: *permutatio*, *Ad Herenn.* IV. xxxiv. 46; Quintilian, *Inst. orat.* IX. ii. 44; Isidore, *Etym.* I. xxxvii. 23; *abbreviatio* (the opposite of *amplificatio*, i.e. conciseness or pointedness, including the various species of *emphasis* or *significatio*), *Ad Herenn.* IV. liii. 67 ff.; *pronominatio* or *antonomasia* (the salient quality in place of the name, *pro nomine*), *Ad Herenn.* IV. xxxi. 42; Geoffrey of Vinsauf, *Poetria nova* 928–34; apostrophe (*exclamatio*), *Ad Herenn.* IV. xv. 22; personification (*conformatio*), ibid. liii. 66; and hyperbole (*superlatio*), ibid. xxxiii. 44. On absolute constructions, Geoffrey of Vinsauf, *Poetria nova*, 701; *Doc.* (ed. Faral), 30 ff. and especially 38.

15. *Ars poetica*, 38–40:

> Sumite materiam vestris, qui scribitis, aequam
> viribus, et versate diu quid ferre recusent,
> quid valeant umeri.

John of Garland, *Paris. poet.* I. 75–82 (Lawler, 6–8).

16. The 'hic opus et labor est' motif is Virgilian (*Aeneid* VI. 129). Horace on poetic discipline, *Ars poet.* 408–18.

17. 'In realtà, la tecnica è in lui una cosa dell'ordine sacrale, è la via del suo esercizio ascetico, indistinguibile dall'ansia di perfezione'

(Contini, ed. *Rime*, x). But see also, on 'internalization', Mengaldo, ed. *DVE*, xlviii: 'Si può ben dire che, diversamente dalle poetiche mediolatine, al centro del discorso dantesco non è l'astratta *eloquentia*, sono i concreti *eloquentes*. Di qui un'interiorizzazione delle norme della poetica, in cui è da vedere una delle ragioni più consistenti della novità del trattato: la validità delle norme non riposa più, alla fine, su criteri oggettivi e prefabbricati, ma sulla dignità dell'esperienza poetante che le fonda con la propria autorità, e quel formalismo è l'altra faccia di questo umanesimo'; and M. Pazzaglia, *Il verso e l'arte della canzone nel De vulgari eloquentia*, Florence, 1967, 111: 'L'arte è per Dante un operare retto della razionalità che definisce e ordina la materia, quella verbale come quella dell'esperienza e dell'emozione, in una limpida armonia di spirito e sensi. Essa esprime, in tal modo, il fondersi nell'atto delle due forme della complessione umana, subordinando alla capacità costruttiva dell'intelletto la sensibilità, senza cancellarla o misconoscerla. L'opera d'arte viene così a riflettere l'*ordo* che la persona è chiamata ad attuare in se stessa e nel mondo.'

18. *Ars. vers.* II. II (ed. Faral 154): 'Ex superficiali ornatu verborum elegantia est in versibus, quando *ex verborum festivitate* versus contrahit venustatem et sibi gratiorem amicat audientiam'; Isidore, *Etym.* II. xxi. 1; Alan of Lille, *Anticlaud.* III. 146–8, etc.

CHAPTER 4

1. In addition to the general histories (Previté-Orton, Davis, Bryce, etc. on the Holy Roman Empire) see on the Frederician period G. Masson, *Frederick II of Hohenstaufen*, London, 1957; T. C. Van Cleve, *The Emperor Frederick II of Hohenstaufen: Immutator Mundi*, Oxford, 1972. On Boniface VIII and Philip IV, R. Scholz, *Die Publizistik zur Zeit Philipps des Schönen und Bonifaz' VIII*, Stuttgart, 1903 (repr. 1962); J. Rivière, *Le Problème de l'église et de l'état au temps de Philippe le Bel*, Louvain, 1926; G. Mollat, *Les Papes d'Avignon*, Paris, 1930 (an English trans. Edinburgh, 1963). On the Henrician period, W. M. Bowsky, *Henry VII in Italy*, Lincoln, Nebraska, 1960; F. Cognasso, *Arrigo VII*, Milan, 1973; and on Ludwig of Bavaria, C. Muller, *Der Kampf Ludwigs des Baiern mit der romischen Curie*, Tübingen, 1879.

2. *MGH*, Legum sectio IV, *Constitutiones*, (ed. I. Schwalm, Hanover-Leipzig, 1906–11), iv. 261 (no. 298). Henry is of the same Gelasian mind; ibid. 340 (no. 391), 343–6 (no. 393), and 395–8 (no. 454)—guarantees of ecclesiastical interests in Italy.

3. *Corpus iuris canonici, Clementinarum,* Bk. II, titt. ix and xi (ed. Richter and Friedberg, Leipzig, 1879–81, part ii, 1147 and 1151).
4. *Corpus iuris canonici, Extravagantes Ioannis Papae XXII,* tit. v. 1211; also in *MGH* v. 340–1 (no. 401).
5. For the dossier on Matteo Visconti, Vatican, Lat. MSS 3936–7, extracts from which are in R. André-Michel, *Mélanges,* Paris, 1920, 184–205. In general G. Mollat, *The Popes at Avignon* (n. 1 above) and more particularly F. Bock, 'Processi di Giovanni XXII contro i Ghibellini', *Archivio della Società Romana di Storia patria* 63 (1940), 129–43; idem, 'I processi di Giovanni XXII contro i Ghibellini delle Marche', *Bullettino dell'Istituto Storico Italiano* 57 (1941), 19–70. Also, L. Fumi, 'Eretici e ribelli nell'Umbria dal 1320–30, studiati su documenti inediti dell'Archivio segreto Vaticano', *Bollettino della R. Dep. di Storia patria per l'Umbria* 3 (1897), 257–85, 429–89; 4 (1898), 221–301, 437–86; 5 (1899), 1–46, 205–425.

    For the indictment of Lewis of Bavaria, *MGH* v. 617 (no. 792).
6. *MGH* v. 646 (no. 824).
7. On Clement V's—the Guasco's—deceit of Henry VII, *Par.* XVII. 82 ('ma pria che 'l Guasco l'alto Arrigo inganni'), while on his simony and corrupt existence generally, *Inf.* XIX. 82–7; on John XXII as the plunderer of Peter's and Paul's vine, *Par.* XVIII. 130–6, and on the Caorsini as usurers, *Inf.* XI. 46–51.
8. For the chronological problem up to and including Vinay, see his edition of the *Monarchia,* Florence, 1950, xxix-xxxviii (Vinay himself, together with D'Ancona, Tocco, Barbi, Pietrobono, Chimenz, and Hardie (in his chronological note to Nicholl's English translation), is for 1312 or 1313); for a more recent bibliographical survey, E. Mongiello, 'Sulla datazione della *Monarchia', Le parole e le idee* 11 (1969), 290–324.
9. Aristotle, *Physics* VII. 3 (247^b17–18); Aquinas, *ScG* I. iv. 4; III. xxxvii. 7; Richard of St Victor, *Ben. maj.* II. vi (*PL* 196, col. 84).
10. Aristotle, *Nic. eth.* VIII. 10 (1160^b2–5); Aquinas ad loc. (n. 1677).
11. Augustine, *De civ. Dei* v. xii; Giles of Rome, *De reg. prin.* (after 1280), I. i. 10 prin. On this line in the canonists (Johannes Andreae Mugellanus), *Mon.*, ed. Vinay, 107 n. 3.
12. Tolomeo of Lucca, *De reg. prin. (c.*1300) II. iv; Engelbert of Admont, *De ortu, prog. et fine Rom. imp. liber (c.*1310), XI.

    D. Thompson, 'Dante's Virtuous Romans', *DS* 96 (1978), 145–62; M. De Matteis, 'Il mito dell'impero romano in Dante: a proposito di *Monarchia* II. i', *LC* 9–10 (1982), 247–56.
13. Cino da Pistoia (Cynus Pistoriensis), *Cod.* VII. 39, *comperit,* no. 2. On Cino as a lawyer and on his political thought: L. Chiappelli,

Vita e opere giuridiche di Cino da Pistoia, Pistoia, 1881; idem, *Nuove ricerche su Cino da Pistoia*, Pistoia, 1911; A. Mocci, *La cultura giuridica di Cino*, Sassari, 1910; C. N. S. Woolf, *Bartolus of Sassoferrato*, Cambridge, 1913; G. M. Monti, *Cino da Pistoia giurista*, Città di Castello, 1924. On Cino and the Donation: D. Maffei, 'Cino da Pistoia e il *Constitutum Constantini*', *Annali della Università di Macerata*, 24 (1960), 95–115; idem, 'Il pensiero di Cino da Pistoia sulla Donazione di Costantino, le sue fonti e il dissenso finale da Dante', *LC* 16 (1987), 119–27. On Cino and Dante, G. Biscaro, 'Cino da Pistoia e Dante', *SM* NS 1 (1928), 492–9. On the Donation itself: B. Nardi, 'La "Donatio Constantini"', in *Nel mondo di Dante*, Rome, 1944, 109–59; idem, 'Dante e il buon Barbarossa', *Alighieri* 7 (1966), 3–27; L. Pietrobono, 'La donazione di Costantino e il peccato originale', *Saggi danteschi*, 2nd edn., Turin, 1954, 167–80; D. Maffei, *La Donazione di Costantino nei giuristi medievali*, Milan, 1964 (repr. 1969).

14. Useful as introductions: J. B. Morrall, *Political Thought in Medieval Times*, London, 1958; G. de Lagarde, *La Naissance de l'esprit laïque au déclin du moyen âge*, 6 vols., 3rd edn., Paris, 1959 (1942); M. J. Wilks, *The Problem of Sovereignty in the Later Middle Ages*, Cambridge, 1963; W. Ullmann, *A History of Political Thought: the Middle Ages*, London (Penguin), 1965; ageing but always useful, O. Gierke, *Political Theories of the Middle Age* (trans. F. W. Maitland), Cambridge, 1900 (reprint 1968) and R. W. and A. J. Carlyle, *A History of Medieval Political Theory in the West*, Edinburgh, 1903–36 (various reprints).
15. Gelasius I, *Tract.* IV. 11 (quoted by Carlyle, i. 190–1).
16. Gregory VII, *Registrum*, iii. 10 (Carlyle, iv. 185).
17. Boniface VIII, *Registrum*, 5382, 'Unam sanctam' (Carlyle, v. 392–3).
18. The key text, in addition to the classic denunciation passages in *Inf.* XIX, *Purg.* XVI, and *Par.* XXVII, is Aquinas's eulogy of Francis and of Franciscan poverty in *Par.* XI.

    U. Cosmo, *L'ultima ascesa*, Bari, 1936, 118–89; B. Nardi, 'Dante profeta', in *Dante e la cultura medievale*, Bari, 1942 (especially 267 ff. on Dante, Olivi, Ubertino, and Franciscan reformism generally). On Joachimite elements in the *Paradiso*: L. Tondelli, *Il libro della figura*, 2nd edn., Turin, 1953; idem, *Da Gioachino a Dante*, Turin, 1944; M. W. Bloomfield, 'Joachim of Fiore', *TR* 13 (1957), 249–311; A. Crocco, *Simbologia gioachimita et simbologia dantesca*, 3rd edn., Naples, 1962; A. Piromalli, *Gioachino da Fiore e Dante*, Ravenna, 1966. The standard work on Joachim in English is M. Reeves, *Joachim of Fiore and the Prophetic Future*, London, 1976.

19. Editions: Giles of Rome, *De ecclesiastica potestate*, ed. R. Scholz, Leipzig, 1961 (1929); James of Viterbo, *De regimine christiano*, ed. H. X. Arquillière, Paris, 1926. Other papal publicists: Henry of Cremona, *De potestate papae*, ed. R. Scholz (n. 1 above), 459 ff. (the *Tractatus brevis* of Augustinus Triumphus is at 486–501); Tolomeo of Lucca, *De regimine principum*, in the Parma edn. of Aquinas, 1863, xvi. 225–91 and ed. J. Perrier in *Thomas Aquinas, Opuscula omnia*, Paris, 1949, i. 220–426. On the imperial and regalian side the *De ortu, progressu et fine Romani imperii* of Engelbert of Admont is in *Politica imperialia*, ed. M. Goldast, Frankfurt, 1614 (ii. 754–72), while the *De regia potestate et papali* of John of Paris (Jean Quidort) is edited by F. Bleienstein, Stuttgart, 1969 (and by J. Leclercq, Paris, 1942).

Secondary: R. Scholz (n. 1 above); J. Rivière (n. 1 above); U. Mariani, 'Scrittori politici medievali', *GD* 29 (1926), 111–40; M. Maccarrone, *Vicarius Christi*, Rome, 1952; idem, 'Potestas directa et potestas indirecta nei teologi del XII e XIII secolo', *Miscellanea Historiae Pontificiae* 18 (1954), 27–47.

More particularly on Giles of Rome: U. Mariani, 'Il *De regimine principum* e le teorie politiche di Egidio Colonna', *GD* 29 (1926), 1–9; idem, *Scrittori politici agostiniani del secolo XIV*, Florence, 1927; on James of Viterbo, in addition to Arquillière (above): U. Mariani, 'Il *De regimine principum* di Giacomo da Viterbo', *GD* 27 (1924), 118 ff.; and on Tolomeo of Lucca, J. Rivière, 'Lucques, Barthelemy de', *Dictionnaire de théologie catholique*, Paris, 1926, 1062–7; T. Silverstein, 'On the genesis of *De Monarchia* II. v', *SP* 13 (1938), 326–49; C. T. Davis, 'Ptolemy of Lucca and the Roman Republic', now in *Dante's Italy and Other Essays*, Philadelphia UP, 1984 (1974), 254–89. On Augustinus Triumphus, M. Wilks, (n. 14 above). On John of Paris, J. Leclercq, *Jean de Paris et l'écclésiologie du XIIIe siècle* (above). On the moderate hierocracy of Remigio de' Girolami see Chapter 2 n. 1 above.

20. In addition to the general histories of medieval political thought (Carlyle, Ullmann, McIlwain, Sabine, etc.), T. Gilby, *Principality and Polity: Aquinas and the Rise of State Theory in the West*, London, 1958 (with bibliography). Important too is Gilby's introduction and commentary on the law and political theory section—Ia IIae 90–7—of the *ST*, vol. 28, London, 1965. J. Catto, 'Ideas and Experience in the Political Thought of Aquinas', *Past and Present* 71 (1976), 3–21. Selections with translations in D. Bigongiari (ed.), *The Political Ideas of St Thomas Aquinas*, New York, 1953 (various reprints), and in A. P. D'Entrèves (ed.), *Aquinas: selected political writings*, Oxford, 1965.

21. Ed. N. Matteini in *Il più antico oppositore di Dante: Guido Vernani da Rimini*, Padua, 1958 (previously, ed. G. Piccini, Florence, 1906). C. Dolcini, 'Guido Vernani e Dante. Note sul testo del *De reprobatione Monarchie*', *LC* 9–10 (1982), 257–62.
22. Nardi, 'Il concetto dell'Impero', in *Saggi di filosofia dantesca*, 2nd edn., Florence, 1967, 231 (see also his 'Di un'aspra critica di fra Guido Vernani a Dante', *Saggi e note di critica dantesca*, Milan–Naples, 1966, 377–85). Gilson, by contrast, on Vernani's perceptiveness, *Dante and Philosophy*, New York, 1963 (1939), 201 and 223–4.
23. Quoted by Nardi in 'Il concetto . . .' (n. 22 above), 256: 'La *Monarchia*, come ha ben detto il Gentile, "è il primo atto della ribellione alla trascendenza scolastica".'
24 'Abbiamo già visto che nella dottrina di Sant'Agostino intorno alla *civitas terrena* vanno distinti due momenti logici: nel primo, lo Stato terreno è concepito come una conseguenza del peccato che ha destato nell'animo umano la libidine del dominare; nell'altro, la *civitas* diventa un mezzo per raggiungere la pace nella giustizia, e la stessa soggezione dell'uomo all'uomo, la *servitus*, è considerata in certo modo come un rimedio naturale contro il peccato stesso da cui trae origine, naturale in quanto "ea lege ordinatur, quae naturalem ordinem conservari iubet, perturbari vetat". In questo senso va intesa appunto la naturalità dello Stato secondo Dante, e non nel senso aristotelico-tomistico. Per lui, il "regimen temporale" in genere, cioè qualunque forma di Stato, e non il solo Impero, è reso necessario dall'imperfezione della natura umana, ed è un rimedio "contra infirmitatem peccati"' (B. Nardi, 'Il concetto . . .' (n. 22 above), 228).
25. Worth noting, over and against a 'second thoughts' or 'later addition' view of the final caveat in III. xv, is that its key emphases are already there earlier on in the book, the 'light of grace' motif in III. iv. 19–20: 'Sed quantum ad melius et virtuosius operandum, recipit aliquid a sole, quia lucem habundantem: qua recepta, virtuosius operatur. Sic ergo dico quod regnum temporale non recipit esse a spiritualibus, nec virtutem que est eius auctoritas, nec etiam operationem simpliciter; sed bene ab eo recipit ut virtuosius operetur per lucem gratie quam in celo et in terra benedictio summi Pontificis infundit illi'; and the 'paternal relationship' motif in III. xi. 6: 'Cum ergo Papa et Imperator sint id quod sunt per quasdam relationes, quia per Papatum et per Imperiatum, que relationes sunt altera sub ambitu paternitatis et altera sub ambitu dominationis, manifestum est quod Papa et Imperator, in quantum huiusmodi, habent reponi sub predicamento relationis, et per consequens reduci ad aliquod existens sub illo genere.' Dante has not, in other words, changed his mind.

26. *Trattatello* (1st red.), 201.
27. 'The Genuineness of the Dedicatory Epistle to Can Grande', in *Studies in Dante*, 3rd Series (Miscellaneous Essays), OUP, 1903 (repr. 1968), 284–374. See now, however, P. Dronke, *Dante and Medieval Latin Traditions*, CUP, 1986, 103–11. Also, F. D'Ovidio, 'L'epistola a Cangrande', in *Studi sulla Divina Commedia*, Milan–Palermo, 1901, 448–85 (against Dante's authorship); F. Torraca, 'L'epistola a Cangrande', in *Studi danteschi*, Naples, 1912, 249–304 (for the letter's authenticity); L. Pietrobono, 'L'epistola a Can Grande', *GD* 40 (1939), 1–51; A. Mancini, 'Nuovi dubbi e ipotesi sull'epistola a Can Grande', *RANL*, ser. vii, 4 (1943), 227–42 (only the first four paragraphs genuine); F. Mazzoni, 'L'epistola a Cangrande', *RANL*, ser. viii, 10 (1955), 157–98; idem, 'Per l'Epistola a Can Grande', in *Studi in onore di Angelo Monteverdi*, Modena, 1959, ii. 498–518; F. Schneider, 'Der Brief an Can Grande', *DDJ* 34–35 (1957), 3–24; B. Nardi, *Il punto sull'epistola a Cangrande (Lectura Dantis Scaligera)*, Florence, 1960 (only the first four paragraphs authentic); idem, 'Osservazioni sul medievale "Accessus ad auctores" in rapporto all'epistola a Cangrande', in *Studi e problemi di critica testuale* (various authors), Bologna, 1961, 273–305.
28. On Dante and allegory: E. Auerbach, 'Figura', in *Studi su Dante*, 2nd edn., Milan, 1967 (1938), 174–221; idem, 'Figurative Texts Illustrating Certain Passages of Dante's *Commedia*', *SP* 21 (1946), 474–89 (also in *Studi su Dante*, 239–60); idem, *Typologische Motive in der mittelalterlichen Literatur*; (Schriften und Vortrage des Petrarca-Instituts, Köln, 2), Krefeld, 1953 (in translation in *American Critical Essays on the 'Divine Comedy'*, ed. R. J. Clements, New York, 1967, 104–13); C. Singleton, *Dante's 'Commedia': Elements of Structure*, Cambridge, Mass. 1954 (reprinted Johns Hopkins UP, 1977); J. Chydenius, *The Typological Problem in Dante: A Study in the History of Medieval Ideas*, Helsinki, 1958; idem, *The Theory of Medieval Symbolism*, Helsinki, 1960; A. C. Charity, *Events and Their Afterlife: the Dialectics of Christian Typology in the Bible and Dante*, CUP, 1966; P. Giannantonio, *Dante e l'allegorismo*, Florence, 1969; R. Hollander, *Allegory in Dante's Commedia*, Princeton, 1969; idem, 'Dante Theologus-Poeta', *DS* 94 (1976), 91–136; J. Pépin, *Dante et la tradition de l'allégorie*, Montreal, 1970; D. Thompson, 'Figure and Allegory in the *Commedia*', *DS* 90 (1972), 1–11; J. A. Scott, 'Dante's Allegory', *RP* 26 (1973), 558–91; P. Armour, 'The Theme of Exodus in the First Two Cantos of the *Purgatorio*', in *Dante Soundings*, Dublin, 1981, 59–99.

On biblical exegesis generally in the Middle Ages: C. Spicq,

Esquisse d'une histoire de l'exégèse latine au moyen âge, Paris, 1944; H. De Lubac, 'A propos de l'allégorie chrétienne', *Recherches de science religieuse* 34 (1947), 180–226 and 47 (1959), 5–43; idem, *Exégèse médiévale. Les quatre sens de l'Ecriture*, Paris, 1959–64; B. Smalley, *The Study of the Bible in the Middle Ages*, 2nd edn., Oxford, 1952; G. W. H. Lampe and K. J. Woollcombe, *Essays in Typology*, London, 1957; J. N. D. Kelly, 'The Bible and the Latin Fathers', in *The Church's Use of the Bible, Past and Present*, ed. D. Nineham, London, 1963, 41–56.

29. e.g. Augustine, *De trin.* xv. ix. 15 (*PL* 42. *c.*1068); Isidore, *Etym.* I. xxxvii. 22; Bede, *De schem. et tropis sac. script.* II. 12 (*PL* 90. *c.*184); Aquinas, *In ep. ad Gal.* IV, lect. 7, etc. For Dante on allegory, or the 'way of the poets' ('lo modo de li poeti'), see *Conv.* II. i. 2–7 — where there is a similar confusion, however, of allegory proper and typology or figuralism.

30. See, however, G. Padoan, 'La "mirabile viš ne" di Dante e l'epistola a Can Grande', in *Il pio Enea, l'empio Ulisse*, Ravenna, 1977, 30–63.

31. On Giovanni del Virgilio, in addition to Martellotti, *ED* s.v.: P. H. Wicksteed and E. G. Gardner, *Dante and Giovanni del Virgilio*, London, 1902; C. Marchesi, 'Le allegorie ovidiane di Giovanni del Virgilio', *SR* 6 (1909), 85–135; G. Albini, 'Giovanni del Virgilio', in *Dante e Bologna*, Bologna, 1922, 45–73; F. Ghisalberti, 'Giovanni del Virgilio espositore delle *Metamorfosi*', *IMU* 4 (1961), 181–200; A. Campana, 'Guido Vacchetta e Giovanni del Virgilio (e Dante)', *RCCM* 7 (1965), 261–3.

On the eclogues as the work of Boccaccio (refuted), G. Padoan in *Studi sul Boccaccio*, i (1963), 517–40 and ii (1964), 475–507 (now in 'Sulla presunta falsificazione delle egloghe dantesche da parte del Boccaccio', in *Il pio Enea* (n. 30 above), 223–51). Boccaccio's *scholia* on the text (MS Mediceo-Laurenziano xxix. 8) are reproduced by Wicksteed and Gardner, 265–311.

32. E. Chiarini, 'I "decem vascula" della prima egloga dantesca', in *Dante e Bologna nei tempi di Dante*, Bologna, 1967, 77–88 — the other possibility being ten Latin eclogues (Novati, D'Ovidio, Parodi, etc.; bibliography in U. Cosmo, *Guida a Dante*, 2nd edn., 1967, 137–9). On the Polyphemus of Bologna, G. Mazzoni, 'Dante e il Polifemo bolognese', in *Almae luces, malae cruces*, Bologna, 1941, 349–72. Biscaro and Cosmo (*Guida a Dante* above, 138) are for Fulcieri da Calboli as Polyphemus (*Purg.* XIV. 58–66).

General Bibliography

I EDITIONS AND TRANSLATIONS

i. Editions

In general, *Opere minori*, Milan–Naples, 1979 (vol. i for the vernacular works, vol. ii for the Latin works).

Rime

M. Barbi et al., *Le opere di Dante* (Società Dantesca Italiana), Florence, 1921, the standard text for K. Foster and P. Boyde, *Dante's Lyric Poetry*, 2 vols., Oxford, 1967 (though numbering differs here) and for G. Contini, *Rime*, 2nd edn., Turin, 1946, with an important introduction. Also M. Barbi and F. Maggini, *Rime della Vita nuova e della giovinezza*, Florence, 1956 and M. Barbi and V. Pernicone, *Rime della maturità e dell'esilio*, Florence, 1969; D. Mattalia, *Le Rime*, Turin, 1943; G. Zonta, *Il Canzoniere*, Turin, 1921.

Detto d'amore and *Fiore*

G. Contini, *Il Fiore e Il Detto d'Amore attribuibili a Dante Alighieri*, Milan, 1984 (supersedes Parodi, 1922).

Vita nuova

M. Barbi, *Edizione Nazionale delle Opere di Dante* (Società Dantesca Italiana), 2nd edn., Florence, 1932 (Milan, 1907), fundamental for Scherillo (1911), Sapegno (1931), Porena and Pazzaglia (1966), De Robertis (1980), and Ciccuto (1984).

Convivio

M. Simonelli, Bologna, 1966; G. Busnelli and G. Vandelli, with an introduction by M. Barbi, 2nd edn., Florence, 1964 (1934–7), with a bibliographical updating by A. E. Quaglio (ii. 415–61).

De vulgari eloquentia

P. V. Mengaldo, *De vulgari eloquentia*, i. Introduzione e testo, Padua, 1968; A. Marigo, 3rd. edn. (with bibliographical updating by P. G. Ricci, 337–82), Florence, 1968 (1957; 1st edn., 1938); B. Panvini, Palermo, 1968. Earlier edns.: Rajna (1896), Bertalot (1917).

Monarchia

P. G. Ricci, Milan, 1965; G. Vinay, Florence, 1950 (the Rostagno text of 1921); E. Moore, with an introduction by W. H. V. Reade, Oxford, 1916 (1894).

Letters, De situ, *and eclogues*

Letters: ed. P. Toynbee, 2nd edn., OUP, 1966 (1920). *De situ*: ed. G. Padoan, Florence, 1968; also V. Biagi, Modena, 1907. Eclogues: ed. E. Cecchini in *Opere minori* (above), superseding Wicksteed and Gardner (1902), Albini (1903, revised by G. B. Pighi, Bologna, 1965), and Pistelli (1921).

ii. Translations

Rime: Foster-Boyde (above), i (facing translations). *Vita nuova* (among the many): M. Musa, *Dante's Vita Nuova*, Indiana UP, 1973; W. Anderson, *The New Life*, London, 1964 (Penguin). *Convivio*: C. Ryan, *Dante: The Banquet*, Saratoga, 1989 (supersedes E. Price Sayer, 1887 and W. W. Jackson, 1909); *De vulgari eloquentia*: P. Wicksteed in the *Latin Works of Dante Alighieri*, London, 1904 (Temple Classics), 1–124; S. Purcell, *Literature in the Vernacular*, Manchester, 1981. *Monarchia*: D. Nicholl, *The Monarchy and Three Political Letters*, London, 1954; P. Wicksteed, op. cit., 125–292 (also has the letters, the eclogues, and the *De situ* in translation).

Marigo's edn. of the *De vulgari eloquentia* and Vinay's edn. of the *Monarchia* have facing Italian translations.

II CRITICISM

i. *Rime*

AUERBACH, E., 'La poesia giovanile di Dante', in *Studi su Dante*, 2nd edn., Milan, 1967 (1929), 23–62.

BARBI, M., *Problemi di critica dantesca*, 2 vols., Florence, 1934 and 1941 (various items).

—— *Studi sul Canzoniere di Dante*, Florence, 1915.

BIONDOLILLO, F., *Le rime amorose di Dante*, Florence, 1960.

BOYDE, P., *Dante's Style in His Lyric Poetry*, CUP, 1971.

—— 'Dante's Lyric Poetry', in *The Mind of Dante* (ed. U. Limentani), CUP, 1965, 79–112.

DE ROBERTIS, D., 'Le Rime di Dante', *NLD* 1 (1966), 285–316.

—— 'Sulle rime', in *Il libro della Vita nuova*, 2nd edn., Florence, 1970, 239–79.

—— 'Sulla cultura giovanile di Dante', *LC* 4 (1973), 229–60.

FIGURELLI, F., *Sulle prime rime di Dante*, Alcamo, 1954.

FOSTER, K., 'Recent Work on Dante's Rime', *Le parole e le idee* 4 (1962), 255–68.

MAGGINI, F., *Dalle 'Rime' alla lirica del Paradiso dantesco*, Florence, 1938 (also in Barbi-Maggini (I. i above), xi–xxxi).

MARTI, M., *Con Dante fra i poeti del suo tempo*, 2nd edn., Lecce, 1971.

MINEO, N., *Dante*, Bari, 1971, 15–31.

MONTANARI, F., *L'esperienza poetica di Dante*, 2nd edn., Florence, 1968.

PARODI, E. G., 'Le Rime di Dante', in *Dante: La vita, le opere, le grandi città dantesche*, Milan, 1921, 53–67.

PAZZAGLIA, M., 'Note sulla metrica delle prime canzoni dantesche', *LS* 3 (1968), 319–31.
PETROCCHI, G., *Politica e letteratura nella vita giovanile di Dante*, Rome, 1974.
SAPEGNO, N., *Storia letteraria del Trecento*, Milan–Naples, 1963.
—— 'Le rime di Dante', *Cultura* NS 9 (1930), 801–17.
ZINGARELLI, N., 'Il Canzoniere di Dante', in *Lectura Dantis. Le opere minori di Dante*, Florence, 1906, 131–62.
ZONTA, G., 'La lirica di Dante', *GSLI*, supp. 19–21 (1922), 45–204.

ii. *Detto d'amore* and *Fiore*

BENEDETTO, L. F., *Il 'Roman de la rose' e la letteratura italiana*, Halle, 1910.
CONTINI, G., 'La questione del "Fiore"', *CS* 13–14 (1965), 768–73.
—— *ED* s.v. *Fiore* (ii. 895–901).
—— 'Un nodo della cultura medievale: la serie "Roman de la rose"—"Fiore"—"Divina Commedia"', in *Un'idea di Dante*, Turin, 1976, 245–85.
—— 'Stilemi siciliani nel "Detto d'amore"', op. cit. 237–43.
PEIRONE, L., *Tra Dante e 'Il Fiore'. Lingua e parola*, Geneva, 1982.
—— *'Il Detto d'Amore' tra 'Il Fiore' e Dante*, Geneva, 1983.
PICONE, M., 'Il "Fiore": struttura profonda e problemi attributivi', *VR* 33 (1974), 145–56.
—— 'Glosse al "Detto d'amore"', *MR* 3 (1976), 394–409.
RICHARDS, E. J., *Dante and the 'Roman de la Rose'*, Tübingen, 1981.
TOOK, J. F., 'Towards an Interpretation of the *Fiore*', *SP* 54 (1979), 500–27.
VALLONE, A., *Dante*, Milan, 1971, 503–27.
VANOSSI, L., *Dante e il 'Roman de la rose'. Saggio sul 'Fiore'*, Florence, 1979.
—— *La teologia poetica del 'Detto d'Amore' dantesco*, Florence, 1974
—— *ED* s.v. *Detto d'amore* (ii. 393–5).

iii. *Vita nuova*

BONFILS TEMPLER, M. DE, *Itinerario di Amore: dialettica di amore e morte nella Vita nuova*, Chapel Hill, 1973.
BRANCA, V. 'La Vita nuova', *CS* 13–14 (1965), 690–7.
—— 'Poetica del rinnovamento e tradizione agiografica nella Vita nuova', in *Miscellanea in onore di I. Siciliano*, Florence, 1966, i. 123–48.
DE ROBERTIS, D., *Il libro della Vita Nuova*, 2nd edn., Florence, 1970.
MARTI, M., 'La Vita Nuova', *Terzo programma* 4 (1965), 73–81.
MINEO, N., *Dante*, Bari, 1971, 31–57.
MONTANARI, F., 'La Vita Nuova', in *L'esperienza poetica di Dante*, 2nd edn., Florence, 1968, 37–102.
MUSA, M., *Dante's Vita Nuova*, Bloomington and London, 1973.
PAZZAGLIA, M., 'La Vita Nuova fra agiografia e letteratura', *LC* 6 (1977), 187–210.

PICONE, M. *Vita nuova e tradizione romanza*, Padua, 1979.
—— 'La Vita nuova e la tradizione poetica', *DS* 95 (1977), 135–47.
PIETROBONO, L., 'La Vita Nuova', *GD* 34 (1933), 113–37 (now in *Saggi danteschi*, Turin, 1954, 1–24).
SAPEGNO, N., 'La vita nuova', in *Pagine di storia letteraria*, Palermo, 1960, 7–28.
SHAW, J. E., *Essays on the Vita Nuova*, Princeton, 1929.
SINGLETON, C. S., *An Essay on the Vita Nuova*, 2nd edn., Cambridge, Mass., 1958 (1949).
SPITZER, L., 'Bermerkungen zu Dantes Vita Nuova', *Publications de la Faculté des Lettres de l'Université d'Istanbul* 2 (1937), 162–208 (Italian translation by C. Scarpati, *Studi italiani*, Milan, 1976, 95–146).

iv. *Convivio*

BRAMBILLA AGENO, F., 'Per l'edizione critica del "Convivio"' in *Atti del Convegno Internazionale di studi danteschi*, Ravenna, 1979, 43–78.
—— 'Nuove proposte per il "Convivio"', *SD* 48 (1971), 121–36.
—— 'Riflessioni sul testo del "Convivio"', *SD* 44 (1967), 85–114.
BERGIN, T., *An Approach to Dante*, London, 1965, 97–152.
DE ROBERTIS, D., 'Il libro della "Vita Nuova" e il libro del "Convivio"', *Studi urbinati* 25 (1951), 5–27.
FOLENA, G., 'La tradizione delle opere di Dante Alighieri', in *Atti del Congresso Internazionale di studi danteschi*, Florence, 1965–6, i. 18–24.
GILSON, E., *Dante and Philosophy*, New York, 1963 (Paris, 1939), 83–161.
MAZZONI, F., *Nota introduttiva a Dante Alighieri, Convivio*, Alpignano, 1965.
—— 'Il "Convivio"', *Terzo programma* 4 (1965), 81–8.
MIGLIORINI FISSI, R., *Dante*, Florence, 1979, 62–71.
MINEO, N., *Dante*, Bari, 1971, 94–110.
PADOAN, G., *Introduzione a Dante*, Florence, 1975, 72–8.
RENUCCI, P., 'Il "Convivio"' in *NLD* 1, Florence, 1966, 327–40.
RICCI, P. G., 'Il Convivio', *CS* 13–14 (1965), 698–704.
SIMONELLI, M., *Materiali per un'edizione critica del 'Convivio' di Dante*, Rome, 1970.
—— *ED* s.v. *Convivio* (ii. 193–204).
—— 'Contributi al testo critico del Convivio', *SD* 30 (1951), 26–46, and 31 (1952), 59–161.

v. *De vulgari eloquentia*

BALDELLI, I., 'Il De vulgari eloquentia' in *Terzo programma* 4 (1965), 113–21.
—— 'Il "De vulgari eloquentia" e la poesia di Dante', *NLD* 8 (1976), 241–58.
BERTONI, G., 'Il De vulgari eloquentia', *AR* 20 (1936), 91–102.
DI CAPUA, A., 'Insegnamenti retorici medievali e dottrine estetiche moderne nel "De vulgari eloquentia"' in *Scritti minori*, Rome, 1959, ii. 252–355.

D'Ovidio, F., 'Sul trattato "De vulgari eloquentia"' in *Opere*, Naples, 1932 (1876), ix. 217–332.
Grayson, C., 'Dante's Theory and Practice of Poetry', in *The World of Dante* (ed. Grayson), OUP, 1980 (1965), 146–63.
—— '"Nobilior est vulgaris": Latin and Vernacular in Dante's Thought', in *Centenary Essays on Dante*, OUP, 1965, 54–76.
Marigo, A., (I. i above), Introduzione, xv–clvi.
Mengaldo, P. V., *ED* s.v. *De Vulgari Eloquentia* (ii. 399–415).
—— *Linguistica e retorica di Dante* (includes the Introduction to his edn., I. i above), Pisa, 1978.
Nencioni, G., 'Dante e la retorica', in *Dante e Bologna nei tempi di Dante*, Bologna, 1967, 91–112.
Pagani, I., *La teoria linguistica di Dante. De vulgari eloquentia: discussioni, scelte, proposte*, Naples, 1982.
Pazzaglia, M., *Il verso e l'arte della canzone nel 'De vulgari eloquentia'*, Florence, 1967.
Rajna, P., 'Il trattato "De vulgari eloquentia"', in *Lectura Dantis. Le opere minori di Dante Alighieri*, Florence, 1906, 195–221.
Schiaffini, A., *Interpretazione del 'De vulgari elquentia' di Dante*, Rome, 1963.
—— *Lettura del 'De vulgari eloquentia' di Dante*, Rome, 1959.
Terracini, B., *Il 'De vulgari eloquentia' e le origini della lingua italiana*, Turin, 1948.

vi. Monarchia

Barbi, M., 'L'ideale politico-religioso di Dante', 'L'Italia nell'ideale politico di Dante', and 'Impero e Chiesa', all in *Problemi fondamentali per un nuovo commento alla Divina Commedia*, Florence, 1956, 49–68, 69–89, 91–114.
Battaglia, F., *Impero, Chiesa e Stati particolari nel pensiero di Dante*, Bologna, 1944.
Chiavacci Leonardi, A. M., 'La *Monarchia* di Dante alla luce della *Commedia*', *SM* 18 (1977), 147–83.
Costanzo, J. F., 'The *De Monarchia* of Dante Alighieri', *Thought* 43 (1968), 87–126.
D'Entrèves, A. P., *Dante as a Political Thinker*, Oxford, 1952.
Davis, C. T., *Dante and the Idea of Rome*, Oxford, 1957.
—— *Dante's Italy and Other Essays*, Philadelphia, 1984 (various essays).
Ercole, F., *Il pensiero politico di Dante*, 2 vols., Milan, 1927–8.
Gilson, E., *Dante and Philosophy*, New York, 1963 (Paris, 1939), 162–224.
Goudet, J., *Dante et la politique*, Paris, 1969.
Herde, P., *Dante als Florentiner Politiker*, Wiesbaden, 1976.
Kelsen, H., *La teoria dello stato in Dante*, Bologna, 1974 (Vienna, 1905).
Lowe, H., 'Dante und das Kaisertum', *Historische Zeitschrift* 190 (1960), 517–62.
Maccarrone, N., 'Il terzo libro della "Monarchia"', *SD* 33 (1955), 5–142.

MACCARRONE, N., 'Papato e Impero nella "Monarchia"', *NLD* 8 (1976), 259–332.
MONTANO, R., 'La *Monarchia* e il pensiero politico di Dante', in *Suggerimenti per una lettura di Dante*, Naples, 1956, 191–219.
MURESU, G., *Dante politico*, Turin, 1979.
NARDI, B., 'Il concetto dell'Impero nello svolgimento del pensiero dantesco' and 'Tre pretese fasi del pensiero politico di Dante', in *Saggi di filosofia dantesca*, 2nd edn., Florence, 1967, 215–310.
—— 'Intorno ad una nuova interpretazione del terzo libro della Monarchia', in *Dal Convivio alla Commedia*, Rome, 1960, 151–313.
—— 'Dante e il buon Barbarossa', *Alighieri* 7 (1966), 3–27.
PARODI, E. G., 'Del concetto dell'Impero in Dante e del suo averroismo', *BSDI* 26 (1919), 106–48.
READE, W. H. V., 'The Political Theory of Dante', introduction to Moore's edn. (I. i. above).
RICCI, P. G., *ED* s.v. (iii. 993–1004).
SOLMI, A., *Il pensiero politico di Dante*, Florence, 1922.
—— 'L'idea imperiale di Dante', in *Studi su Dante*, Milan, 1944, 1–31.
SUMNER, B. H., 'Dante and the "Regnum Italicum"', *MAEV* 1 (1932).
VASOLI, C., 'La dottrina politica di Dante', in *La filosofia medievale*, Milan, 1961, 406–10.
VINAY, G., *Interpretazione della Monarchia*, Florence, 1962.

vii. Letters

FRUGONI, A., 'Le epistole', in *Dante nella critica d'oggi* (ed. U. Bosco), Florence, 1965, 739–48.
PARODI, E. G., 'Intorno al testo delle epistole di Dante e al Cursus', in *Lingua e letteratura*, Venice, 1957 (1912), ii. 399–442.
TORRACA, F., 'Le lettere di Dante', in *Nuovi studi danteschi*, Naples, 1921, 137–79.
TOYNBEE, P., (I. i above), Introduction, xiii–liv (with bibliography on particular letters, 253–7).

viii. De situ

FRECCERO, J., 'Satan's Fall and the Quaestio de aqua et terra', *IT* 38 (1961), 99–115.
GHISALBERTI, A., 'La cosmologia nel Duecento e Dante', *LC* 13 (1984), 33–48.
MAZZONI, F., 'Il punto sulla Quaestio de aqua et terra', *SD* 39 (1962), 39–84.
—— 'La Quaestio de aqua et terra', *SD* 34 (1957), 163–204.
MOORE, E., 'The Genuineness of the *Quaestio de aqua et terra*', in *Studies in Dante*, Oxford, 1968 (1899), ii. 303–57.
NARDI, B., 'La caduta di Lucifero e l'autenticità della Quaestio de aqua et terra', *Lectura Dantis Romana*, Turin, 1959.
PADOAN, G., 'La Quaestio de aqua et terra', in *Dante nella critica di oggi* (ed. U. Bosco), Florence, 1965, 758–67 (with bibliography).
PASTORE STOCCHI, M., in the *ED* s.v. (iv. 761–65).

ix. Eclogues

ALBINI, G., 'Le egloghe', in *Lectura Dantis; le opere minori di Dante Alighieri*, Florence, 1906, 259–82.
BATTISTI, C., 'Le egloghe dantesche', *SD* 33 (1955–6), 61–111.
BOLSANI, E. and VALGIMIGLI, M., *La corrispondenza poetica di Dante Alighieri e Giovanni del Virgilio*, Florence, 1963.
MARIGO, A., *Il classicismo virgiliano nelle egloghe di Dante*, Padua, 1910.
MARTELLOTTI, G., in the *ED* s.v. (ii. 644–6).
—— 'La riscoperta dello stile bucolico: da Dante al Boccaccio', in *Dante e la cultura veneta*, Florence, 1966, 335–46.
MAZZONI, F., 'Le egloghe', in *Dante minore*, Florence, 1965, 79–114.
PADOAN, G., *Introduzione a Dante*, Florence, 1975, 113–18.
REGGIO, G., *Le egloghe di Dante*, Florence, 1969.
SALSANO, F., 'Le egloghe' in *Dante nella critica d'oggi*, Florence, 1965, 749–57.
VECCHI, G., 'Giovanni del Virgilio e Dante. La polemica tra latino e volgare', in *Dante e Bologna nei tempi di Dante*, Bologna, 1967, 61–76.
WICKSTEED, P. H. and GARDNER, E. G., *Dante and Giovanni del Virgilio*, London, 1902.

Index

abbreviatio 141
active life (ideal of) 106–11, 115, 166
additio 8, 17, 145
Aelred of Rievaulx 51, 202, 203
Alan of Lille 43, 214
Albert the Great 96, 97, 100, 109, 112, 209
allegory 179–81, 219–20 (in the moral *rime*) 63–6, 87–8 (in scripture) 155–6, 180–1, 220
Ambrose, St 175
amplificatio 213
analogy 98, 204 (in the *Vita nuova*) 54–5 (in the moral *rime*) 65–6
anaphora 23
Andreas Capellanus 1–2, 35, 193, 206, 207
angel, *see* intelligence
antonomasia 141, 213
apostrophe 141
Aristotle (Aristotelianism) 51, 59, 94–8, 105–9, 117, 121, 128, 157, 166, 168, 169, 171, 181, 208, 209, 210, 211, 215
Arnaut Daniel 8, 66–7, 205
ars cantionis, *see* canzone
ars dictaminis 140, 213
astrology (science of) 87
astronomy, *see* cosmology
Augustine (Augustinians, Augustinianism) 97, 106, 109, 119–20, 128, 169–70, 182, 201, 202, 203, 204, 207, 208, 210, 211, 215, 218, 220
Augustus (Caesar) 152, 153, 169
Averroes (Averroists, Averroism) 97, 112–17, 166–7
Avicenna 97
Aquinas, *see* Thomas

Battifolle (Countess) 177
Beatrice 21–2, 53, 55, 57, 59, 203
Bede (the Venerable) 175, 220
Bernard of Clairvaux 51, 53, 182, 208

Bernardus Silvestris 43
Boccaccio 173, 220
Boethius (*Consolation of Philosophy*) 43, 52, 61, 88, 91, 96, 201, 202, 208, 209, 212
Boethius of Dacia 111, 209
Bonagiunta da Lucca 3, 11
Bonaventure, St 96, 198, 202
Boniface VIII (Pope) 147, 161, 214, 216
Brunetto Latini 33, 135, 202, 203, 205, 211

Can Grande della Scala 179, 185, 219
cantus divisio 136
canzone (structure of) 136–8
Capellanus, *see* Andreas
capfinidas 6
Cato 153
Cavalcanti, Guido 4–5, 9–10, 16–17, 18–19, 20, 21, 27, 45, 48–50, 67, 79, 85–6, 131, 196–7, 206
Cecco Angiolieri 31, 69
Charlemagne 156
Charles of Valois 141
Christ 153–4, 156, 203 (in the *Vita nuova*) 54–5
Cicero 51, 61, 88, 92, 134, 140, 202, 203, 207, 208, 209, 211, 212
Cino da Pistoia 4, 38, 77–9, 131, 207, 215–16
Clement V (Pope) 147, 148–9, 174–5, 214
comic style 41, 69–70, 144
Constantine 159
constructio 140–2, 145
convenientia (literary principle of) 8, 135, 145
conversio 15, 23, 24, 25
cosmology (Dante's system of) 86–7, 99–101
cursus 140, 141, 212–13

Damascene (John of Damascus) 175
Dante da Maiano 12–15, 144, 197–8

INDEX

Decretalists 155
dialectic (science of) 87
Dionysius the Areopagite 96, 181–2
dolce stil novo (stilnovisti, stilnovismo) 4–7, 9–11, 21–9, 135, 195–6
Dominicans 61 (*see too* Guido Vernani, Remigio de' Girolami, Santa Maria Novella)
Donation of Constantine 77, 156–7

earthly paradise (image of) 158
Eden 119, 128–9, 169
Egidius Colonna, *see* Giles of Rome
emanation (emanationism) 86–7, 95–6
emperor (role of) 90–1, 157–73 *passim*
emphasis 23–4, 213
Empyrean 86, 87, 100–1, 183
Engelbert of Admont 153, 215, 217
epanalepsis 24
ethic (Dante's idealist) 101–5 (Dante's Peripatetic) 106–9
ethics (science of) 88, 109–10
exile (in Dante) 73–7, 81–2, 132–4, 177–8
exornatio 8, 17, 60

figuralism, *see* typology
fin'amor 32, 34
foll'amor 44
Forese Donati 69–70, 144, 260
forma tractandi (and *forma tractatus*) 179
Franciscans (Franciscan, Friars Minor) 61, 97, 109, 149, 162, 216 *see also* Olivi, Ubertino, Santa Croce
Frederick I (Barbarossa) 160
Frederick II 90, 133–4, 160, 163, 214
Frederick of Austria 149
frons (canzone) 136
Frontinus 141

gabbo (episode in *Vita nuova*) 50, 202
Geoffrey of Vinsauf 134, 137, 145, 212, 213
geometry (science of) 87
Gelasius I (Pope) 160, 214, 216
genera dicendi 138–40
Ghibellines 148, 149
Giacomo da Lentini 2, 6, 22, 32
Giles of Rome (Egidius Colonna) 162–3, 212, 217
Giovanni del Virgilio 185–7, 220
Giraut de Bornelh 8

grammar (science of) 87
grammatica, see Latin
Gregory, St 175
Gregory VII (Pope) 160, 216
Guelphs 153, 155, 177–8
Guido da Polenta 179
Guillaume de Lorris 2, 30–1, 200
Guinizzelli, Guido 4, 6–7, 22, 32, 63, 197
Guittone d'Arezzo (Guittoniani, Guittonianism) 3–4, 9, 73, 141–2, 194–5

heavenly paradise (image of) 158
Henry VII (of Luxemburg) 147–9, 174–5, 185, 214, 215
Henry of Cremona 217
hierocracy (papalism) 159–66
Horace 129, 211, 213
human law, see *lex humana*
hyperbole, see *superlatio*

idealism (philosophical) 94 (in the *Convivio*) 98–105
improperium 47
intelligence (separate substance, angel, mover) 7, 86–7, 101, 102–4
inversion, see *prolepsis*
irony (difficult figure of) 141 (in the *Fiore*) 41 (in *Doglia mi reca*) 75–6
Isidore of Seville 139, 198, 213, 214, 220

James of Viterbo 163–4, 217
Jean de Meun 2, 31, 36–43 *passim*, 52, 91, 199–200, 201, 202, 208
Joachim of Fiore (Joachimism) 216
John XXII (Pope) 147, 175
John of Garland 135, 139, 145, 212
John of Jandun 97
John of Paris (Jean Quidort) 217
John the Baptist 54
justice 73–5, 151–2

language (as a principle of individuality) 126–7 (evolutionary nature of) 128–9 (utilitarian function of) 126 (earliest history of) 129–31
Langue d'oc 129
Langue d'oil 129
Lapo Gianni 20, 131, 198
Latin (as the language of

INDEX

philosophy) 83–5, 123–4 (as common currency) 130
Latin Averroists (radical Aristotelians) 97, 209
Lewis IV of Bavaria 149, 215
lex humana (human law, positive law) 73, 90–1 (in Aquinas) 164
lexis (in poetry) 25, 27, 74, 137–8, 140
Liber de causis (*Book of Causes*) 101, 209
Liberalitade (in *Doglia mi reca*) 75–7 (in the *Convivio*) 84–5, 121–2
Lingua di sì 129
Livy 141
love (perception of in Sicilians, Tuscans, and *stilnovisti*) 1–7 (in the *Fiore*) 34–43 (in the *Vita nuova*) 43–58 *passim* (in the *montanina*) 77–81 (in the *Convivio*) 88–9 (as the formal cause of philosophy) 89 (as primarily in God) 89–90
Lucan 141
Lucifer 184
Lucy, St 112

Manfred 133
Mary (Virgin) 112
Matthew of Vendôme 134, 137–8, 145, 214
Meo de' Tolomei 69
metaphysics (science of) 87, 109–10
metaphor 141 (in the *Vita nuova*) 59–60 (in the *petrose*) 67–9
monopsychism (Averroes) 113–14, 167
Moroello Malaspina 79
music (science of) 87 (and poetry) 135–8

Neoplatonism 95–6 (*see too* Idealism)
Niccolò da Prato (Cardinal) 177
nobility (*gentilezza*) 6–7, 71–3, 90–3
numerus carminum et sillabarum (in poetry) 137

Oberto (and Guido) da Romena 177
oda continua 66
Olivi, Pier di Giovanni 205, 216, *see also* Franciscans
ordo artificialis 140
Orlandi, Guido 10
ornatus facilis 141
Orosius 141
Ovid (and Ovidianism) 1, 10, 141, 192–3

papalism, *see* hierocracy
partium habitudo (canzone) 136–7
Paul, St 53, 96
permutatio, see irony
pes (canzone) 136
Peter, St 148, 154, 156
Peter of Blois 51, 202
Petrarch 185
Philip IV (the Fair) 147, 161, 214
philosophy 88–90 (as love of wisdom) 89 (in relation to theology) 114, 158, 188
physics (science of) 87
Pierre d'Alvernh 8
planctus (lyric genre) 47
Plato (Platonists) 94–5; *see also* Idealism
Plotinus 95–6
Pope (role of) 157–73 *passim*
Porphyry 96
positive law, see *lex humana*
possible intellect 113–14, 151, 166–7
poverty (ideal of) 156–7, 162
praise style (*stilo de la loda*) 21–9, 62–3
primum mobile 86, 100, 101, 183
Proclus 96
prolepsis (inversion) 23, 140, 141
pronominatio, see *antonomasia*
Protonoè (first mind) 86, 94, 100, 183
Pseudo-Dionysius, *see* Dionysius the Areopagite
Ptolemy 86, 100
Pythagoras (Pythagoreanism) 87, 96

Quintilian 8, 134, 140, 212, 213

Remigio de' Girolami 204
Replicatio (repetition) 23, 28; see also *anaphora*, *epanalepsis*
rhetoric (classical and medieval) 8, 134 (science of) 88 (and poetry) 138–42
rhyme 67, 70, 74, 137
Richard of St Victor 53, 56, 182, 201, 202, 203, 204, 207, 211
rima composta 13
rima equivoca 13
Robert of Naples (of Anjou) 148, 175
Romans (as rulers) 153–4
Rome (providential nature of) 117–19
Rustico di Filippo 31, 69

Santa Croce 205
Santa Maria Novella 204
Santo Spirito 205

Scripture (hierocratic interpretation of) 155–6 (typological interpretation of) 180–1, 220
Seneca 202, 203, 208, 209, 211
sestina doppia 67, 206
Sicilian (Sicilians, *scuola siciliana*) 2–3, 8–9, 63, 193
Siger of Brabant 97, 111
sirima (sirma, canzone) 136
Solomon 114
soul (genesis of) 91, 113–14
stanza (structure of), *see* canzone
Statius 141
stilnovisti (stil novo), see *dolce stil novo*
superlatio (hyperbole) 23, 213
syntax (in poetry) 24, 25, 67–9, 74

tenzone (lyric genre) 22 (the Forese *tenzone*) 69–70
Teruccio di Manetto Donati 178
theology (science of) 88 (in relation to philosophy) 109–10, 114, 188
Thomas Aquinas, St 73, 98, 102, 105, 109, 112, 151, 164–6, 169, 202, 207, 208, 209, 210, 211, 212, 216, 217, 220
Tolomeo di Lucca 153, 215, 217
tragic style (*stilus tragicus*) 140, 144
transumptio, see metaphor

trobar clus (and *trobar leu*, closed and open styles) 8, 16
Tuscan (as vernacular standard) 131
Tuscan school (*scuola toscana*) 3–4, 9, 194
typology 180–1

Ubertino da Casale 205, 216; *see also* Franciscans
Uguccione 212
Unam Sanctam (of Boniface VIII) 161–2

Valla, Lorenzo 156
vernacular (Dante's use of for philosophy) 83–5, 121, 124–8 (as a principle of consciousness) 123–8
Vernani, Guido 167–8, 218
verso chiave 136
versus (canzone) 136
Villani, Giovanni 173
Virgil 111, 112, 117–19, 150, 210
Visconti, Matteo 215
vituperatio (lyric genre) 16
vulgare illustre 128–34, 142–4

William of Moerbecke 96
wisdom (as primarily in God) 89–90 (book of) 207